# Safe food handling

## A training guide for managers of food service establishments

Michael Jacob

*Environmental Health Officer*
*Department of Health*
*London, United Kingdom*

World Health Organization
Geneva
1989

Reprinted 1992

WHO Library Cataloguing in Publication Data

Jacob Michael
Safe food handling: a training guide for managers of food service establishments.

1. Food contamination – prevention and control
2. Food handling – methods 3. Food services – standards
I. Title

ISBN 92 4 154245 4 (NLM Classification: WA 701)



The designations employed and the presentation of the material in this publication do not imply the expression of any opinion whatsoever on the part of the Secretariat of the World Health Organization concerning the legal status of any country, territory, city or area or of its authorities, or concerning the delimitation of its frontiers or boundaries.

The mention of specific companies or of certain manufacturers' products does not imply that they are endorsed or recommended by the World Health Organization in preference to others of a similar nature that are not mentioned. Errors and omissions excepted, the names of proprietary products are distinguished by initial capital letters.

The author alone is responsible for the views expressed in this publication.

TYPESET IN INDIA
PRINTED IN ENGLAND

88/7646 Macmillan/Clays—4500
92/9205—Clays/GCW—1500

# Contents

# Preface

Incidents of foodborne disease, especially those widely publicized in the media, highlight the problems that need to be overcome to achieve safe preparation, handling, storage and distribution of food. They also act as a constant reminder that foodborne disease occurs even where formal national or local government controls are in force. Routine inspections of food service establishments may help to ensure that food is prepared in a clean environment, but often cannot control other factors that contribute to foodborne disease. Carrying out daily inspections is neither practical nor effective.

The recent emergence, in some developed countries, of foodborne illness associated with foods not previously implicated, e.g., salmonellae in eggs and *Listeria* in chilled foods, indicates that contamination of raw products can be a problem. However, on a worldwide basis, most such illness is caused by foods that have been mishandled or mistreated during preparation. One of the most effective preventive measures to deal with foodborne illness is thus to educate food-handling personnel in safe practices.

This guide has been designed to help overcome the problems associated with educating food handlers. It is written for anyone involved in managing a food service establishment, or supervising food handlers, and is appropriate for use in hotel and catering management courses. Its emphasis is on bacterial foodborne diseases as these diseases can readily be prevented by the adoption of safe food handling techniques.

*

* *

The author is grateful to all those people who reviewed the early drafts of this guide, and whose comments and suggestions were most helpful in the preparation of the final version.

# How to use this guide

Parts I–III of this guide give details on how food contamination occurs and how to prevent it by employing a variety of measures including safe food handling. Part IV gives guidance to help managers organize this information into a training course for food handlers.

The guide has been written taking into consideration the wide range of educational backgrounds of managers and supervisors in food service establishments. The most crucial points from each chapter are collected in boxes labelled 'Important training points'. Managers with only a basic level of education should concentrate on understanding and remembering these points. Managers with a higher level of education will find the more detailed information in the chapters useful. It is the 'Important training points' that should be stressed in the food handlers' training course. The illustrations throughout the guide can be used to clarify points during the training course.

## Important training points

Chapter 1

# Introduction

## Foodborne illness – outlining the problem

A foodborne illness generally involves a disturbance of the gastrointestinal tract, with abdominal pain, diarrhoea, and sometimes vomiting. Illness is caused by eating food containing a significant amount of harmful (pathogenic) bacteria, or the toxic products of their growth. The illness may affect an individual, one or two members of a family or other close group, or many people. The symptoms may be mild, lasting only for a few hours, or serious, lasting for days, weeks or months, and needing intensive treatment. In vulnerable groups, such as infants and the elderly, the illness is likely to be more severe.

### Public health

Foodborne illnesses continue to be a major public health problem in the developed and developing worlds alike. Statistics tend to underestimate the number of cases of foodborne illness because not everyone affected visits a doctor, and doctors may not report all the cases they treat to the appropriate health authority. Some cases may not be recognized as foodborne illness.

### Factors contributing to foodborne illness

Current statistics for foodborne illness in various industrialized countries show that up to 60% of cases may be caused by poor food handling techniques, and by contaminated food served in food service establishments. No valid data are available for most developing countries, but there is reason to believe that they have similar problems. Bacterial contamination of food can, however, be eliminated by hygienic handling. Effective cooking, followed by appropriate hot or cold storage of the cooked foods are the principal safety factors with poultry and other meats.

Some public health measures can themselves affect food safety. For example, the use of sewage for irrigation can contaminate crops and lead to parasitic and other infections of both humans and food animals. Potentially toxic chemicals, for example pesticides used in agriculture, can find their way into foods. The indiscriminate use of insecticides and rodenticides in kitchens also creates hazards. Harmful natural toxins occur in food, and toxic metals and compounds can find their way into food from utensils, food containers or work surfaces. Viruses, yeasts, and moulds can also be responsible for foodborne illness. The factors that contribute to foodborne illness are summarized in Fig. 1.

Fig. 1. Summary of factors that contribute to foodborne illness.

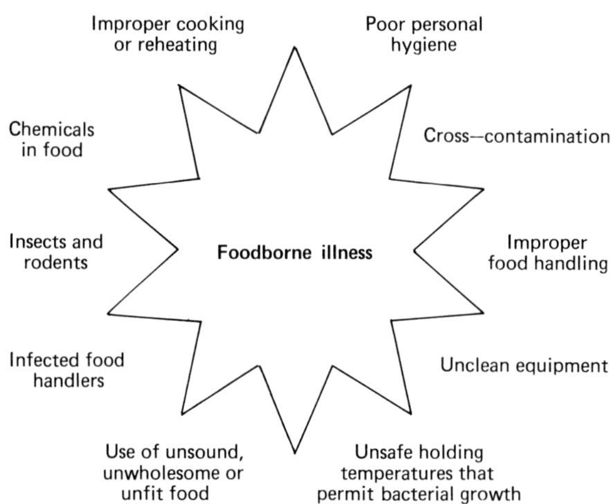


## Economic aspects

Foodborne illness can be responsible for much economic loss and hard work for those involved in the food service industry. Working days and income may also be lost for the customers affected by the illness. Contaminated food has to be withdrawn from sale and discarded. There may be adverse publicity on television or radio, or in the newspapers. This publicity can lead to economic loss, closure, law suits or prosecution. These effects can be out of all proportion to the illness and discomfort suffered by affected customers. It is hard to build up a good reputation in the food service industry, but easy to lose it.

Some countries miss out on potential income from tourism because of the prevalence of serious foodborne illnesses, such as typhoid and dysentery.

Investigations of outbreaks of foodborne illness can also be very time-consuming and expensive. The investigating authorities may spend many hours tracing the social contacts of affected people, assessing attack rates in groups of affected people, and collecting and examining samples of food, stools, and urine from affected people or suspected carriers.

## Consumer service

Food service establishments perform a public service. In many countries meals eaten away from home are an important feature of leisure activities, particularly in tourist areas. Clean and hygienic presentation of food is expected by customers.

Clean surroundings are likely to promote good standards of behaviour among food handlers. People tend to respond to their environment, so it is likely that workers will stay longer in their jobs if their place of work is clean and pleasant. The tourism and leisure industries benefit particularly from high standards of hygiene and cleanliness in food service establishments.

## The effects of foodborne illness on digestion

To understand how foodborne pathogens act on the human body, a basic understanding of the digestive process is needed.

Food may be defined as any solid or liquid that, when swallowed, supplies the body with energy for growth or functioning. For this energy to be released, food must be broken down into its various parts by the process of digestion. Digestion takes place in the alimentary canal, which is the tube that passes from the mouth to the anus. The mechanical action of the teeth, the muscular movement of the alimentary canal, and the chemical reactions caused by the digestive juices all help to achieve this. The complete process of breakdown is called digestion.

The process of digestion converts nutrients from the food into a form that can be absorbed by the body. Absorption involves transferring these nutrients through the walls of the alimentary canal and into the blood.

The main method of propelling food along the alimentary canal is peristalsis. The walls of the canal contract in waves and propel the food onwards. If poisonous or irritant substances enter the stomach, they induce a reverse peristalsis which, combined with contractions of the abdominal muscles and diaphragm, produces vomiting. If the irritant substances are not removed by vomiting, they pass into the intestine and produce contraction, pain and diarrhoea. Different types of foodborne pathogens may cause different body reactions, but generally the clinical features of foodborne illness are diarrhoea, abdominal pain, vomiting and possibly fever. There may be associated symptoms of nausea, prostration, and dehydration. Dehydration is especially common, and life-threatening, in infants and children.

## Introduction to microbiology

In learning about food safety there is a need for an elementary knowledge of microbiology, which is the study of all forms of plant and animal life too small to be seen by the naked eye. In food microbiology, attention is restricted to four groups of organisms—bacteria, moulds, yeasts, and viruses.

As well as being present in humans, these organisms are found in soil, air, water, and often in, or on, our food. Foodborne illness is most often caused by bacteria and viruses.

## Bibliography

CHARLES, R. H. G. *Mass catering*. Copenhagen, WHO Regional Office for Europe, 1983 (European Series No. 15).

FREEDMAN, B. *Sanitarians' handbook*, 4th ed. New Orleans, Peerless Publishing Company, 1977.

LONGREE, K. *Quantity food sanitation*. New York, Inter-Science Publishers, 1967.

RIEMANN, H. & BRYAN, F. L. *Food-borne infections and intoxications*, 2nd ed. New York, Academic Press, 1979.

WHO Technical Report Series, No. 705, 1984 (*The role of food safety in health and development*: report of a Joint FAO/WHO Expert Committee on Food Safety).

# PART I

# CAUSES OF FOOD CONTAMINATION

# Chapter 2
# Bacteria

A bacterium consists of only one cell—it is unicellular. Bacteria are so small that individually they cannot be seen without a microscope. They may be as small as 0.0005 mm, and clusters of a thousand or more are only just visible to the naked eye; 50 000 placed side by side may measure barely 25 mm. Bacteria are the most common cause of foodborne illness.

## Bacterial growth

Bacteria consume food as a source of energy, and for cell growth. A bacterium must absorb food through its cell wall. To do this it requires a suitable environment.

### Temperature

Bacteria grow best within a certain temperature range. They are classified into three groups, depending on which temperature range they prefer (see Fig. 2).

- *Psychrophilic (cold-liking bacteria)*
  Growth range 0–25 °C.
  Optimum temperature 20–25 °C.
- *Mesophilic (middle-liking bacteria)*
  Growth range 20–45 °C.
  Optimum temperature 30–37 °C.
- *Thermophilic (heat-liking bacteria)*
  Growth range 45–70 °C.
  Optimum temperature 50–55 °C.

The species that cause disease and infection in humans grow best at body temperature (37 °C) and are therefore mesophilic. The ones that cause food spoilage in the refrigerator are psychrophilic. If the temperature is below their normal growth range, bacteria will usually not grow. However, they may not be killed by this low temperature and will often start growing again when favourable temperature conditions return.

On the other hand, if bacteria are heated above their normal temperature range for a significant period of time, they will be killed (see Fig. 3). For any given species to be killed a specific combination of time and temperature is needed.

### Time

When bacteria find suitable conditions, reproductive growth can occur. Bacteria reproduce by dividing themselves into two equal

Fig. 2. Important temperatures in food safety.

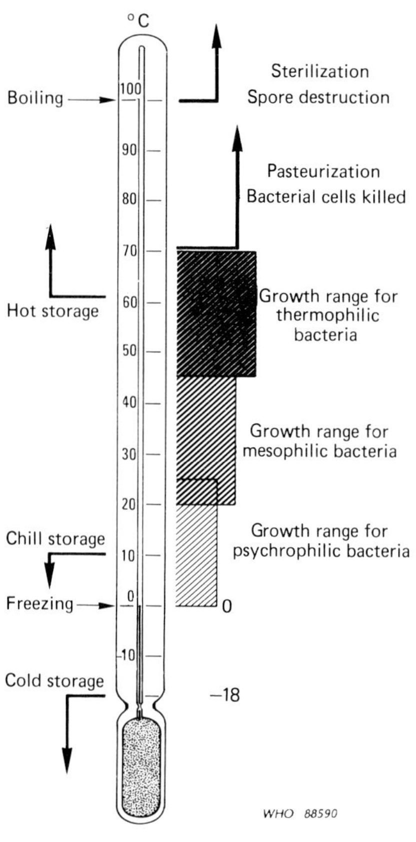


portions. Under suitable conditions of environment and temperature, division occurs once every 20 to 30 minutes. Thus under favourable conditions for continuous growth one cell could give rise to over 17 million in 8 hours, and 1 billion ($10^9$) in 10 hours (see Fig. 4).

## Moisture

Bacterial cells are composed of approximately 80% water. Water is an essential requirement for them. However bacteria cannot use water if it is combined with solids, for example salt and sugar. Concentrated solutions, for example 200 g/litre salt solution, do not generally support the growth of bacteria.

Fig. 3. Control of pathogenic bacteria by temperature.

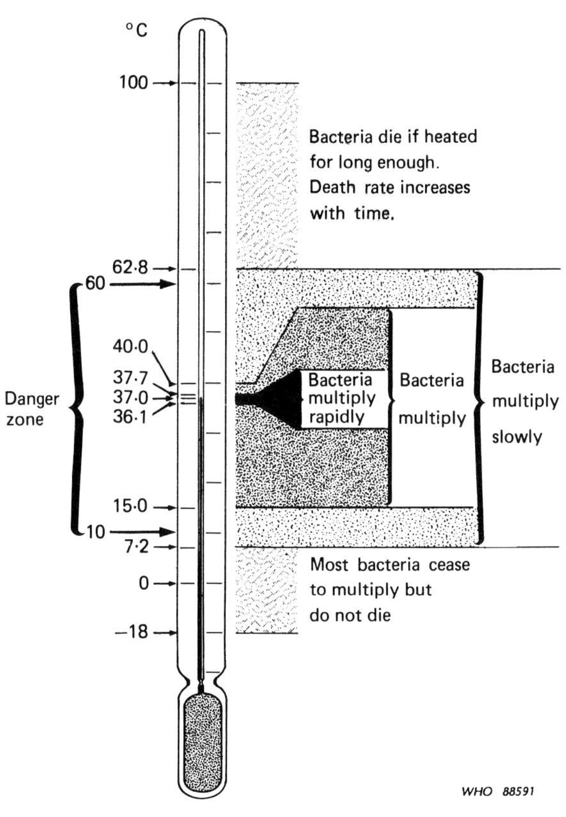


## Oxygen

Some bacteria will grow only if oxygen is present (aerobic bacteria), others only if it is absent (anaerobic bacteria). Others, called facultative anaerobes, can live without oxygen, but prefer an environment where oxygen is present.

## pH

The acidity or alkalinity of a substance is measured on the pH scale. This refers to the hydrogen-ion concentration of a substance. A pH of 7 is neutral (for example water), a pH below 7 is acid, and a pH above 7 is alkaline. Most bacteria prefer a slightly alkaline pH of between 7.2 and 7.6, although some are able to withstand more extreme conditions. For

Fig. 4. Multiplication of bacteria under favourable conditions.

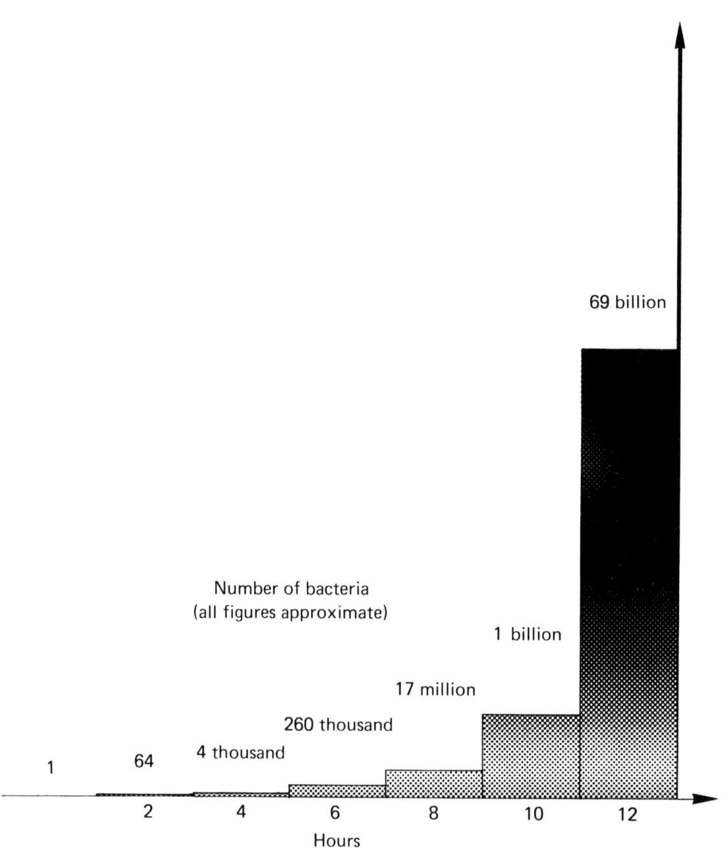


example, the lactic acid bacteria responsible for turning milk sour, and bacteria used in cheese production can tolerate an acidic pH as low as 4.

### Light

Bacteria usually grow best in darkness, although this is not a necessity. Ultraviolet light is lethal to them and may be used in some sterilization procedures.

## Bacterial spores

Most bacteria die in the absence of nutrients or under unfavourable environmental conditions. Certain bacteria, however, develop resistant spores, that have outer protective shells to withstand the unfavourable conditions. Spore-producing bacteria are of particular significance in foodborne disease because they may be able to survive normal cooking temperatures.

## Bacterial toxins

Many pathogenic bacteria (i.e., those that cause disease) produce complex enzymes that will destroy protein and tissue. These enzymes are known as toxins. Some toxins, such as *Staphylococcus* toxin, are heat-resistant. This makes them highly dangerous in foods, as cooking may not destroy them.

## Sampling for the presence of bacteria

Bacteriological examinations can be used to find out if a person is infected, or a piece of food contaminated. Specimens for examination may be collected in a number of ways. For example, a sterile cotton wool swab on the end of a wooden stick may be used to collect a small amount of infected material from a wound. Faeces may be collected in sterile glass or plastic containers. Samples of food material, or swabs from food preparation areas may also be collected.

Foodborne pathogenic bacteria do not usually alter the appearance, taste or smell of food so generally it is impossible to know whether food is dangerously contaminated or not, without bacteriological examination.

Bacteria that break down protein so that there is noticeable spoilage, with putrefaction and smell, do not always cause illness, but if obvious spoilage is present it is best not to allow the food to be eaten.

## How bacteria cause foodborne illness

For foodborne disease to occur, one of the following three events must take place:

- either bacteria must be present on, or in, the contaminated food in sufficient numbers or concentrations to survive the growing period, harvesting, storage, and processing;

- bacteria on, or in, foods must multiply and reach sufficient quantities, or produce toxins in sufficient quantities to cause illness;
- bacteria must enter the food preparation area on, or in, raw foods, and be transferred to workers' hands, or to equipment and surfaces, which, if inadequately washed, will contaminate other foods.

In all these cases, for illness to occur, sufficient quantities of the contaminated food, containing enough bacteria or toxin to overcome a person's resistance to the disease, must be consumed. The number or concentration of bacteria needed to cause illness is known as the Minimum Infective Dose (MID). Ingestion of foods contaminated to this level may result in individual or sporadic cases of illness as well as outbreaks. Whether or not outbreaks are detected depends on the number of people who consume the contaminated food, and whether subsequent illness is reported and investigated.

If a number of pathogenic bacteria are consumed, but not enough to cause illness, an infected individual may become a carrier and may contaminate other foods that he or she touches. The number of organisms required to produce illness depends on the virulence of the organism, the age and general health of the person, and probably many other factors as well.

As an example, in an extensive international outbreak of foodborne disease due to salmonellae involving chocolate candy, illness was caused by the consumption of well under one salmonella cell per gram. Usually, in foodborne illness caused by salmonellae, 10 000 or more organisms per gram are needed to cause illness.

In the case of dysentery due to shigellae, where the most common means of transmission is person-to-person, either directly or through contamination of the environment, the infective dose of shigellae may be in the range of 200–10 000 cells.

Infants, elderly or undernourished people, and people already affected by other illness are more susceptible to foodborne illness than healthy adults, and a dose lower than the normal MID could cause illness, and even fatalities among them.

Following the consumption of infective bacteria or a toxin, there will be an incubation period during which the bacteria multiply and/or a reaction occurs in the gastrointestinal tract, before illness occurs. The duration of the illness varies with the type of bacteria or toxin, but in most cases the illness lasts from one to three days. Patients with salmonella infection normally recover from the gastrointestinal symptoms within 2 or 3 days but may feel generally unwell for a week or more afterwards.

Identification of illnesses of bacterial origin is usually made on the basis of (1) isolation of the causative agent, (2) the incubation period, (3) the clinical features, and (4) the duration of the illness. The most common bacterial foodborne illnesses are described below and summarized in Table 1.

## Bacteria that cause foodborne diseases

### Salmonellae

One of the commonest species of salmonellae is *Salmonella typhimurium*, but other species, such as *S. enteritidis*, which has been found in eggs, may also be involved in causing disease. The bacteria frequently live in animals that are either clinically or subclinically infected. The bacteria reach food directly or indirectly in a variety of ways for example from animal excreta, human excreta, or water polluted by sewage. In the kitchen they may be transferred from raw to cooked food by hands, surfaces or utensils and other equipment.

The incubation period is from 6 to 72 hours, usually 12 to 36 hours, and the symptoms are diarrhoea, abdominal pain, vomiting and fever. The illness lasts for several days and even healthy adults may take up to 21 days to recover fully.

In recent years there have been numerous incidents of salmonellosis arising from the consumption of poultry meat and eggs. Contamination of poultry is sometimes due to infection of the live birds by feed contaminated with salmonellae.

A small number of healthy, well-nourished people may become infected with the bacteria but not show any symptoms. If the bacteria are not completely eliminated from the body, such people become carriers of the organism and salmonellae may continue to be excreted in their faeces. Carriers should not normally be employed in handling food that is ready to eat or easily open to contamination.

Table 1. Summary of bacterial foodborne diseases.

| Causative agent | Incubation time (hours) | Clinical features | Duration of illness |
|---|---|---|---|
| salmonellae (usually *Salmonella typhimurium*) | 6–72 (usually 12–36) | Diarrhoea, abdominal pain, vomiting, and fever | Several days; up to 3 weeks. |
| *Staphylococcus aureus* | 1–6 (usually 2–4) | Nausea, vomiting, abdominal pain, prostration, dehydration, and subnormal temperatures | 1–2 days |
| *Clostridium perfringens* | 8–22 (usually 12–18) | Diarrhoea and abdominal pain. Vomiting is rare. | 1–2 days |
| *Clostridium botulinum* | 12–96 (usually 18–36) | Dizziness, headache, tiredness and double vision, accompanied by dryness of the mouth and throat, followed by an inability to speak due to paralysis of the throat muscles. Death often occurs as a result of paralysis of the respiratory centres. | 3–7 days in fatal cases. Otherwise months or years to recovery. |

| | | | |
|---|---|---|---|
| *Bacillus cereus* | 6–16 (classical outbreaks) | Acute diarrhoea and occasional vomiting. | Generally no longer than 24 hours |
| | 1–6 | An acute attack of nausea and vomiting with some diarrhoea. | |
| *Escherichia coli* | 12–72 (usually 12–24) | Abdominal pain, fever, vomiting, and diarrhoea that may be prolonged with blood and mucus in the stools. | 3–5 days |
| *Vibrio parahaemolyticus* | 12–24 | Abdominal pain, vomiting and diarrhoea, leading to dehydration and fever. | 1–7 days |
| *Campylobacter* species | 72–120 | Fever, followed by persistent diarrhoea with foul-smelling and often bile-stained stools. | 1–10 days |

## Staphylococci

Foodborne illness due to staphylococci, the most common of which is *Staphylococcus aureus*, depends on the presence of sufficient toxin in the food. The incubation period is short and the symptoms appear 1–6 hours after the food has been eaten. Symptoms are nausea, vomiting, abdominal pain, prostration, dehydration and subnormal temperature. Symptoms do not last for long—usually from less than one day to two days. A healthy person can frequently harbour staphylococci on the skin, or in the nose or throat. A septic lesion may also be a source of the organism. Most outbreaks are caused by the direct contamination of cooked food by hands contaminated with secretions from the nose, mouth, wounds or skin. Often cooked food is contaminated with staphylococci by being handled while it is still warm. Staphylococci, finding the subsequent storage conditions favourable, multiply and produce toxin.

The staphylococcus itself is fairly readily destroyed by the heat of pasteurization or normal cooking, but these procedures will not destroy the toxin. To ensure its destruction, boiling for a period that would virtually disintegrate the food is required.

Frequent hand washing by people handling food is a very important factor in preventing contamination risks from staphylococci. People with infected wounds on their hands or arms should not handle food before the infection is healed.

## *Clostridium perfringens*

This is an anaerobe that will, however, tolerate a small amount of oxygen. It is commonly found in excreta from humans and animals, in raw meat and poultry, and in other foods, including dehydrated products. It can survive heat and dehydration by means of its spores which remain dormant for long periods in food, soil and dust.

Illness occurs after eating food contaminated with *Clostridium perfringens* bacteria that have developed from spores surviving cooking. They may be activated to germinate by the heat. Illness results from the production of toxin when the bacteria revert to the spore form, in the intestine. An effective dose of toxin is not formed in the food before it is eaten.

The symptoms, which occur 8 to 22 hours after the food has been eaten, include diarrhoea and abdominal pain. Vomiting is rare. Symptoms last from less than one day to two days and are followed by rapid recovery for healthy people.

Raw meat and poultry are common sources of *Clostridium perfringens*. Most outbreaks of poisoning occur in canteens, hospitals, schools, hotels, and other institutions where meat and poultry dishes are often precooked, cooled slowly, and then reheated. The slow cooling is one of the most critical events leading to illness.

## *Clostridium botulinum*

This organism is anaerobic and is particularly dangerous as it can form spores in canned and vacuum-packed foods, where air is absent. Botulism is caused by the toxin produced as *Clostridium botulinum* grows in food. The toxin is lethal even in small doses. It affects the nervous system, and often causes fatal illness even in previously healthy, strong individuals. The spores are resistant to heat and survive boiling and high temperature. The toxin, however, is sensitive to heat and, in pure form, is destroyed by boiling. It may, however, be protected when mixed with protein and other material in food. The incubation period varies from 12 to 96 hours (usually 18 to 36 hours), the symptoms are dizziness, headache, tiredness and double vision, accompanied by dryness of the mouth and throat, followed by an inability to speak due to paralysis of the throat muscles. Death often occurs by paralysis of the respiratory centres. Unless adequately treated, a third of patients die 3 to 7 days after the onset of the illness. Even with adequate treatment recovery is slow, taking perhaps months and sometimes years.

High-protein foods such as meat and fish are susceptible to contamination. In these foods blackening may occur and gas may be produced, making the food visibly inedible. However, in slightly acid conditions with little protein present, no blackening occurs and the gas is produced in limited quantities only. In these cases contamination may be unnoticed. Meat, fish and vegetables, canned or bottled at home, have been the main cause of outbreaks of botulism. Dangers occur when there is inadequate heat treatment or inadequate preservation with acids or salts. The integrity of commercially canned foods is generally extremely high, but cases of botulism have occurred in people after consuming commercially produced canned foods. There have been two recent outbreaks where commercially canned salmon contaminated with *Clostridium botulinum E* caused the deaths of three people, and severe illness in three others. The outbreaks were thought to be caused by damage to cans allowing recontamination of the contents after the heating and packing operations were completed.

### *Bacillus cereus*

This is a sporing bacillus (rod-shaped bacterium) which occurs in the soil and is a common contaminant of cereals and other foods. Some spores will survive cooking and germinate into bacilli that grow and produce toxin. Storage of warm, cooked food for long periods in moist conditions encourages the growth and division of the organisms to large numbers and the consequent formation of toxin in amounts able to cause illness. The incubation period varies from 6 to 16 hours and the onset of symptoms may be sudden with acute diarrhoea and occasional vomiting. A wide variety of food, particularly boiled rice and cornflour sauce, has been associated with outbreaks.

A different set of symptoms can also be caused by *Bacillus cereus*. These symptoms are an acute attack of nausea and vomiting, with some diarrhoea. The incubation time in these cases is only 1 to 6 hours compared with the 6 to 16 hours in a classical outbreak.

Therefore, *Bacillus cereus* is an organism that can produce two clinically different types of foodborne disease, one closely resembling that of *Clostridium perfringens* and the other similar to staphylococcal food poisoning.

### *Escherichia coli (E. coli)*

Some strains of *E. coli* can cause acute gastroenteritis affecting adults and children. Outbreaks, and sporadic cases in hospitals and maternity homes, are often associated with fatalities in babies. It now seems that many, if not most, outbreaks of traveller's diarrhoea are caused by certain types of *E. coli*. Infants and children are often infected by direct faecal–oral spread, person-to-person contact, and also by eating contaminated food. Large doses of enteropathogenic *E. coli* in foods are thought to be responsible for the disease in adults.

The bacteria may be present in many raw foods and are readily passed to cooked foods by means of contaminated hands, surfaces, containers and other equipment. They may be present in water, and human excreta may also play a direct part in the spread of infection during epidemics. In conditions where standards of hygiene are low and there is a general lack of sanitation, substantial numbers of people, both adults and children, may be at risk from foods contaminated with *E. coli*.

Flies landing on food or equipment surfaces after having had contact with open drains can also spread the bacteria.

The incubation period is 12–72 hours. The symptoms are abdominal pain, fever, vomiting and diarrhoea, which may be prolonged, with blood and mucus in the stools. Symptoms rarely last for longer than 3–5 days.

### *Vibrio parahaemolyticus*

Outbreaks of illness associated with this organism, reported mostly from Japan, usually involve the consumption of raw or cooked seafood. In warm weather the organism can be isolated from fish, shellfish and other seafoods, and from coastal waters. *Vibrio parahaemolyticus* has been found in fish landed from northwestern European, Mediterranean, Adriatic, American and Australasian waters, as well as around eastern Asia. As in the transmission of salmonellae, raw or insufficiently heated foods present the most danger. The incubation period is usually from 12 to 24 hours. Symptoms of the illness are abdominal pain, vomiting and diarrhoea leading to dehydration and fever. The illness lasts from 1 to 7 days.

### *Campylobacter*

The incidence of foodborne disease caused by *Campylobacter* species may be greater than recognized previously. Poultry meat, milk, and water have been implicated in incidents. Birds and household pets, including dogs, have been identified as sources of the organism. Food handlers may be infected by handling raw animal products, especially poultry. Person-to-person spread has also been demonstrated, but is less common. Multiplication of the organism does not occur within food, and it is destroyed by normal cooking and pasteurization temperatures.

Campylobacter causes an acute enteric disease with a usual incubation period of 3–5 days. Symptoms last for 1–10 days. Onset may be sudden with abdominal cramps followed by passage of foul-smelling and often bile-stained stools. The diarrhoea may persist for 1–4 days and is sometimes preceded by a period of fever of a few hours to several days.

In industrialized countries the disease is mainly acquired by eating food derived from infected animals, or cross-contaminated during processing (see Fig. 5). The transmission in developing countries appears to be predominantly through improper handling of food, and through water contaminated by faecal material from infected people and animals, or by direct or indirect contact with them or their faeces. Environmental contamination appears to be important in the spread of infection in developing countries, particularly among the poorer sectors of the population where domestic cattle, fowl and people are often housed together (see Fig. 5).

### *Listeria monocytogenes*

Listeriosis is caused by a bacterium, *Listeria monocytogenes*, a widely distributed environmental contaminant whose primary means of transmission to humans is through contamination of foodstuffs at any point in the food chain. Listeriosis is a relatively uncommon disease, producing acute mild fever in non-susceptible individuals. However, in pregnant women, fetuses, newborn children, and people whose immune system is compromised, the disease can be much more severe and the case-fatality rate is high. Available data suggest that the incubation period is between one and several weeks.

Several major food commodities have been implicated in transmission, including milk and dairy products, meat (especially raw meat products), poultry and its products, vegetables, salads, and seafood. Unlike most other foodborne pathogens, *L. monocytogenes* is able to multiply at refrigeration temperatures of 4–6 °C. Correct application of listericidal processes, such as pasteurization, cooking or irradiation, will rid foods of the organism. The risk of contamination can also be reduced by adherence to hygienic practices.

## Communicable bacterial diseases

Nausea, vomiting, abdominal pain, and diarrhoea are clinical features of several communicable diseases that can be transmitted through food. The main diseases that occur as localized outbreaks are outlined below and summarized in Table 2.

### Typhoid and paratyphoid fevers

These are diseases of the whole body, which may show gastrointestinal symptoms and may be foodborne. The incubation

Fig. 5. Transmission of *Campylobacter.*

a) Common animal reservoirs and sources of *Campylobacter.* (The gull is feeding on a rubbish tip and the sheep has just given birth to a campylobacter-infected stillborn lamb.)
b) Direct transmission to occupational groups, for example farmers, butchers and poultry processors.
c) Direct transmission by person-to-person contact.
d) Indirect transmission through milk, water and meat.
e) Cross-contamination to other foods.

Adapted from: *Report of the WHO Consultation on Veterinary Public Health Aspects for the Prevention and Control of Campylobacter Infections.* Moscow, 1984. (Unpublished WHO document, VPH/CDD/COS/84.1).
Copies of this document can be obtained by writing to: Veterinary Public Health, Division of Communicable Diseases, World Health Organization, 1211 Geneva 27, Switzerland.

Table 2. Summary of bacterial diseases that can be transmitted through food.

| Disease | Incubation time (days) | Clinical features | Duration of illness |
|---|---|---|---|
| Typhoid fever and paratyphoid fever | 7–21 | Continued fever, headache and cough. Enlargement of the spleen, and rose-coloured spots on the trunk. Constipation is more common than diarrhoea. | 3–4 weeks |
| Cholera | up to 5 (usually 2–3) | Sudden onset of profuse, watery stools; vomiting; rapid dehydration; acidosis; and circulatory collapse. | 1–2 days if treated. Up to 7 days if untreated. |
| Shigellosis (bacillary dysentry) | 1–7 (usually 1–3) | Diarrhoea, fever, nausea, and sometimes vomiting and cramps. Stools may contain blood, mucus, and pus. | Average 4–7 days, but can last several weeks. |

period (7–21 days) is longer than usual for foodborne diseases. Typhoid and paratyphoid fevers are caused by bacteria that are acquired almost exclusively from human sources. The disease lasts for 3 to 4 weeks, and has a slow recovery time. Some patients remain as carriers for a long time, and occasionally people become carriers without ever showing symptoms, making them very difficult to trace.

The symptoms are continued fever, headache and cough; enlargement of the spleen and rose-coloured spots on the trunk. Constipation is more common than diarrhoea.

There is a substantial risk of typhoid in areas where there is poor general sanitation and no water purification. Water supplies in areas with nonexistent or defective drainage systems can be directly contaminated with typhoid organisms. All water used in food service establishments for preparing food and for washing food utensils and working surfaces should be of drinking-water quality.

## Cholera

This is a serious acute intestinal disease caused by *Vibrio cholerae*, and characterized by a sudden onset of profuse, watery stools, vomiting, rapid dehydration, acidosis, and circulatory collapse. Humans are the reservoir of infection and the illness is transmitted through the consumption of water contaminated by faeces from patients or carriers, or through food that has been contaminated by water, soiled hands, or flies. Shellfish harvested from water contaminated by faeces play a particular role in the epidemiology of cholera. The incubation period is from a few hours, to five days (usually 2–3 days).

## Shigellosis (bacillary dysentery)

This is an acute bacterial disease characterized by diarrhoea accompanied by fever, nausea, and sometimes vomiting and cramps. Stools may contain blood, mucus and pus. Outbreaks are common in overcrowded conditions and where sanitation is poor. Humans are the reservoir of infection and the illness is usually passed on by direct or indirect faecal–oral transmission from a patient or a carrier. Food handlers failing to wash contaminated hands after defecation may spread infection by direct contamination of food or food preparation surfaces. The incubation period is from 1 to 7 days (usually 1–3 days).

In one outbreak of shigellosis, residents of an old people's home became ill after eating shrimp. The shrimp was contaminated prior to importation, or became contaminated during thawing and repacking for distribution. Fourteen people died. Thirteen of them were over 75 years old, and were more susceptible to the organism because of their age and frailty.

## Bacteria
## Important training points

- Heat-resistant bacterial spores and toxins in food present dangers of foodborne illness.
- To grow and multiply bacteria need:

  Warmth
    Blood heat is the most favourable for the growth of foodborne pathogens.

  Time
    Although bacteria multiply rapidly, time is required for the numbers to reach a level sufficient to cause illness.

  Moisture
    Like all living organisms bacteria require water (and food) to thrive.

- In bacterial foodborne diseases, a minimum infective dose (MID) of bacteria or toxin is needed to cause illness. Handling food so that the number of bacteria is kept below the minimum infective dose is essential.
- If a person consumes less than the minimum infective dose he or she may become a carrier.
- Salmonellae are a very common cause of foodborne illness, mainly because they are easily spread thorough the environment to people, domestic animals and wildlife.
- Staphylococcal contamination of food can be prevented by careful personal hygiene, and thorough hand-washing by food handlers.
- The spores of *Clostridium perfringens* cause foodborne illness because they survive reheating in previously prepared meat dishes.
- Campylobacter foodborne illness can be caused by contaminated food or water, and occurs where hygiene standards are poor.

Chapter 3

# Other food contaminants

## Viruses

Viruses are even smaller than bacteria. Some have capsules or outer coats that protect them. Viruses are recorded as causing outbreaks of intestinal illness even though they cannot multiply in foods, only in certain living tissues. Their spread from the hands of human carriers, and from water to food is important. The presence of viruses in foods, particularly shellfish grown in sewage-polluted water, may be significant in causing illness. Even if routine microbiological examination of food and water does not reveal a significant number of bacteria, the food may still contain pathogenic viruses. There is some evidence that viral foodborne diseases might be more widespread than so far assumed, but it is reassuring that the viruses implicated will not survive the normal heat treatment applied in routine cooking procedures.

### Hepatitis A

Epidemiological information shows that the hepatitis A virus is mainly spread by food. However, the disease's long incubation time (15–50 days (usually 28–30 days)) makes it difficult to investigate outbreaks. The symptoms are fever, malaise, nausea and abdominal discomfort, followed by jaundice.

Shellfish from polluted areas, water, fruits and vegetables contaminated by faeces, and various types of salad prepared under unhygienic conditions have all been involved in outbreaks.

## Chemicals

Chemical contamination of food is often caused by careless storage of pesticides, detergents, or sterilizing agents, leading to spillage or leakage. Also if these substances are kept in unmarked containers they may be confused with food ingredients. Containers used for pesticides or detergents and then for food, without decontamination in between, have also caused illness.

When chemicals are bought in bulk and decanted into smaller containers for convenience, it is important that these containers are clearly labelled and stored well away from all food. As well as directly contaminating food, chemicals may taint it. Chemical odours, particularly from phenolic disinfectants, may get into food and make it inedible even when it is packed.

Cases have been reported of chemical contamination of food from the containers or cooking utensils used to prepare it, for example zinc poisoning from galvanized iron containers used for stewing acid fruit, and antimony poisoning of acidic drinks kept in enamelware of inferior quality.

Other substances, mainly metals, that may contaminate food include cadmium, copper, arsenic, lead, mercury, and their compounds. Other chemical contaminants of importance are pesticide residues picked up by animals during feeding, and antibiotics and other drugs given to food animals to prevent disease or induce growth.

Chemical contamination of food can be widespread and serious. Incidents of mercury contamination of fish, contaminated cooking oil, and vinylchloride contamination from plastic packaging have been well publicized.

An incident where a large number of people became ill through eating bread made from flour contaminated by a chemical hardener for epoxy resin illustrates the dangers of storing raw food ingredients near chemicals. In this case the chemical hardener was carried in the same vehicle as sacks of flour, and spillage of the chemical occurred.

From the food handler's point of view, detecting chemical contamination in food arriving in the catering premises is part of general quality control (see page 108). Ensuring that only reliable sources of food are used, asking for guarantees from suppliers, and checking containers of food on arrival for signs of extraneous matter or taint, provide some safeguards.

## Parasites

Parasitic infections of food are difficult to investigate as little is known about the infective dose required, or the precise mode of transfer of the infective agent to the individual. Contamination may occur from hand to food or directly from polluted water. Problems arise in many parts of the world where raw or under-cooked meat or fish are eaten, and people drink untreated water, or use it in food preparation. Using food only from reliable sources, preparing food using safe water supplies, and using adequate cooking and refrigeration temperatures should reduce the risks of serving food containing parasites.

### Giardiasis

This disease is caused by the flagellated protozoan *Giardia lamblia* (see Fig. 6). Cysts of this organism can be absorbed by

Fig. 6. Modes of transmission of giardiasis.

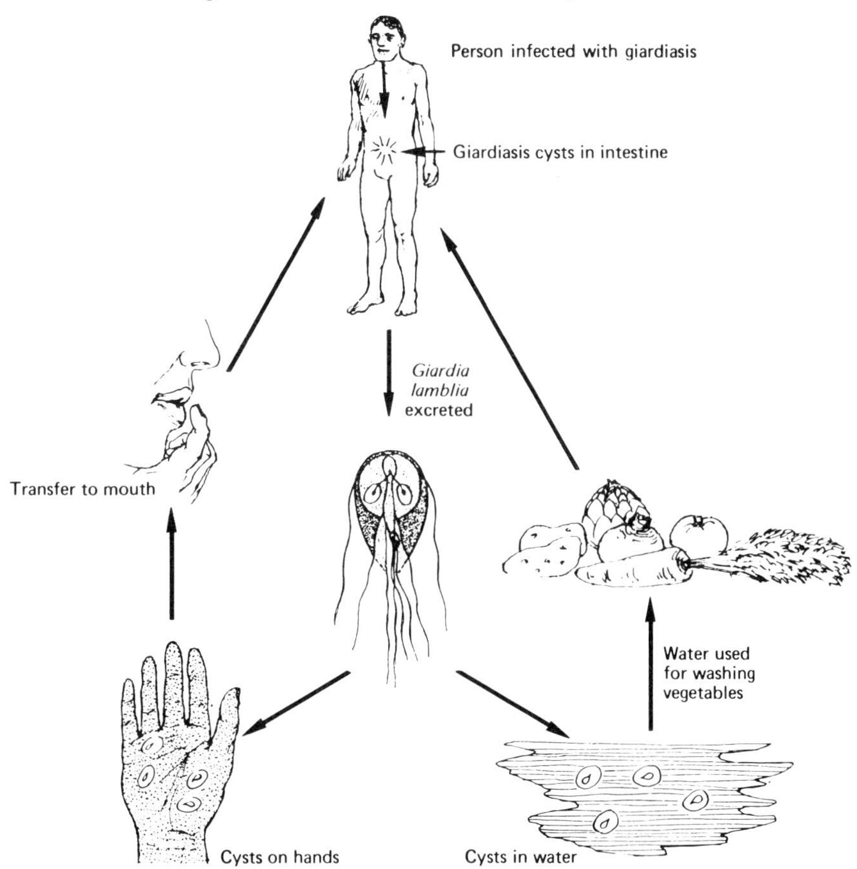


humans. *Giardia* cysts penetrating the intestinal walls cause discomfort, nausea and loose stools. People may become carriers without having symptoms. Children are infected more often than adults. The disease occurs more often in areas of poor sanitation, and in institutions. Giardiasis cysts are excreted in faeces, and are spread by faeces entering water that may later be used to wash vegetables. There may also be hand-to-mouth transfer of cysts in an infected person. The incubation period of giardiasis may be anything from 5 to 25 days, or longer. The main preventive measures are personal hygiene, the sanitary disposal of faeces, and the protection of public water supplies against faecal contamination.

### Trichinellosis

This disease is caused by the migration through the body of larvae of the helminth (worm) *Trichinella spiralis*. People are infected by eating raw or insufficiently cooked meat, chiefly pork and pork products, containing the trichinella cysts. In the small intestine, larvae develop into mature adults and mate. Female worms then produce larvae which penetrate the intestinal wall and eventually enter the bloodstream. The larvae encyst themselves in skeletal muscle.

The reservoirs of infection are swine, dogs and cats, and many wild animals including foxes, bears, marine mammals, and rats.

Symptoms include fever, retinal haemorrhage, diarrhoea, muscle soreness and pain, skin lesions, and prostration. The incubation period varies between 5 and 45 days. The disease is not transmitted directly from person to person. Animal hosts remain infective for months and meat from the animals can cause disease for long periods unless treated to kill the larvae. The modes of transmission of trichinellosis are illustrated in Fig. 7.

Preventive measures are inspection of meat in the slaughter house and adequate cooking for pork and pork products. Low temperatures are also effective in killing trichinellae (for example holding at −25 °C, or lower, for 10 days).

## Natural food contaminants

Certain plants are poisonous to humans. Generally these plants are eaten by accident—the plant, fruit, or berry being mistaken for a similar edible variety. Certain fungi are also extremely poisonous or produce toxic substances. They must be identified accurately before they are eaten, in case they are mistaken for edible varieties, for example mushrooms.

Ergot is a fungal disease of cereal crops, especially rye. The cereal grains become poisonous if affected with fungus. Flour made from these grains will also be poisonous. The flour is normally discoloured, but this may not be noticed. If the flour is eaten, the toxin may affect the nervous system, sometimes with fatal results.

A large number of mould products have adverse effects on animals, and there is growing evidence that these toxic products can also affect people. At least 200 different types of mould, when growing in certain foods, under suitable conditions, form substances that are toxic when eaten. Aflatoxins and other myco-

Fig. 7. Modes of transmission of trichinellosis.

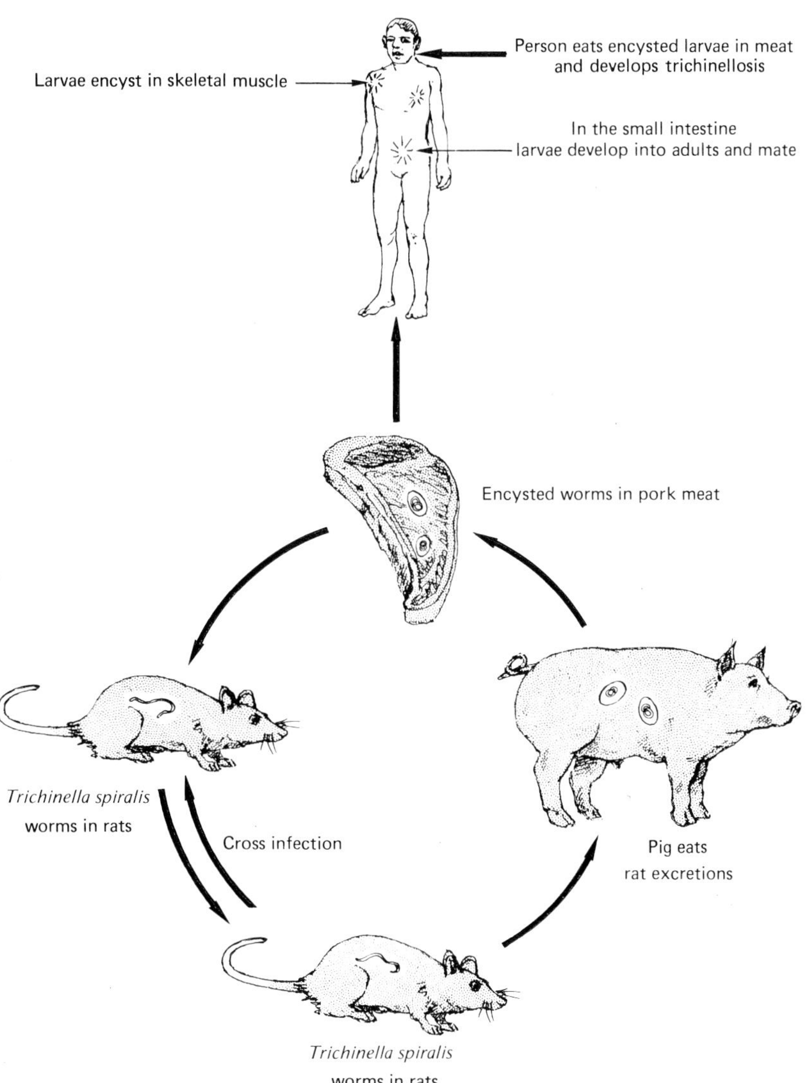


toxins seem capable of causing serious illness, with effects on the liver and other organs.

Good storage practices are essential to prevent mould developing on foods such as cereal flours, rye bread, nuts and fruit juices.

Other examples of contaminants occurring naturally in basic foods are—solanine, a toxic alkaloidal substance in green potatoes, hydrogen cyanide, given off from unripe almonds, and erucic acid in rape seed oil. These natural contaminants may cause illness under certain circumstances.

There is more risk of illness if a toxin-containing food makes up a large part of the diet of an individual or community. People with a more varied and balanced diet are less likely to be affected.

Fish and shellfish may contain a number of naturally occurring toxins that can become poisonous when eaten. Symptoms of fish toxin poisonings include nausea, vomiting, abdominal cramps, diarrhoea, flushes, headache, urticaria (transient, raised, red or pale patches on the skin) and a burning sensation in the mouth. Many species of tropical fish contain neurotoxins that are not destroyed by ordinary methods of cooking. For example, in Japan the fugu fish, a species of puffer fish, must be properly prepared by trained and licensed cooks. Even then it regularly causes illness and some deaths. Some species of barracuda have been involved in ciguatoxin poisoning.

Fish of the order Scombroidea, which includes tuna and mackerel, have caused scombroid poisoning. This occurs when fish is stored at too high a temperature, and may be caused by the formation of excessive amounts of histamine and related compounds in the fish. This problem is reduced if the fish is stored below 8 °C and consumed as fresh as possible.

Molluscan shellfish, such as mussels, clams, and scallops, may cause poisoning if they have been feeding on a certain type of dinoflagellate (unicellular, biflagellate organism present in plankton) that produces a neurotoxin. These dinoflagellates contaminate the sea to such an extent in some parts of the world that they may discolour areas of water and be called 'red tides'. Food or health authorities monitor shellfish in these areas to ensure that those with dangerous toxin levels are not eaten.

Poisoning due to naturally occurring contaminants in some foods cannot necessarily be prevented by good food safety measures. Management in food service establishments must be aware of such foods so that they can exercise appropriate quality control over raw materials entering their establishments.

## Other food contaminants
## Important training points

- Chemical contamination of food is much less common than bacterial contamination. When it occurs in food establishments it is due mainly to carelessness in handling chemicals.
- The possibility of large-scale chemical contamination of food means that management must be aware of possible dangers with particular foods, and be careful when buying them.
- There should be reduced risks of parasites in food if:
  Water supplies are reliable.
  Community sanitation involves proper disposal of faeces.
  Personal hygiene among food handlers is good.
  Meat is thoroughly cooked.
- Naturally occurring contaminants in food are not removed by safe food handling practices. The answer is to be aware of problems that may arise with local supplies of any particular food.
- To prevent decomposition and the risk of scombroid poisoning, keep fresh fish below 8 °C. Cook and serve as soon as possible after delivery. Storage and display of fish in ice not only keeps it safe but also enhances its appearance.

Chapter 4

# Incidents of foodborne illness

The following are examples of incidents of foodborne illness, giving the faults in food handling that lead to outbreaks. The case studies illustrate, among other things, the role of cross-contamination.

## Salmonellosis

### Example 1

An outbreak of salmonellosis occurred affecting 19 people, 18 of whom had eaten at one restaurant; the 19th was a cook at the restaurant. Symptoms were diarrhoea (in 100% of cases), abdominal cramps (in 100% of cases), fever and vomiting starting between 8 and 39 hours (with an average of 17 hours) after eating. The illness lasted from 1 to 17 days. Salmonellae were isolated from 10 of 11 stool specimens submitted by people with symptoms, and from the cook who had had diarrhoeal illness during the same 3-day period.

All 18 customers had eaten cold turkey in either sandwiches or turkey salad. Investigation revealed that all the turkey served in the restaurant was cooked on the premises by roasting to an internal temperature of 57–60 °C rather than the 74 °C required by local regulations. Cooked turkeys were then refrigerated until served in sandwiches or salad. Hot turkey dishes were prepared by reheating sliced meat to a temperature of over 74 °C. In this case, reheating apparently provided sufficient additional cooking since only cold turkey dishes were implicated in the outbreak. The cook, although producing positive stool specimens, was probably not the source of the outbreak. He ate at the restaurant and probably became infected there.

#### *Faults*

- Inadequate initial cooking temperature, insufficient to destroy salmonellae.
- Possibly storage for too long after cooking, in changing temperatures favourable to multiplication of bacteria.

### Example 2

Four people ate a late-night meal of curried chicken and chips bought at a cafeteria. All became ill 24–36 hours later, with vomiting and diarrhoea. Faecal samples were taken from all those affected. Salmonellae were isolated from three of them as well as from the remnants of the food retrieved from the dustbin. A search for other cases revealed that 15 more people were ill. They had all bought similar meals from the same cafeteria. Onset of symptoms was 16–96 hours after eating the curry, and salmonellae were isolated from all 15 people.

Investigation revealed that food handling techniques and cleaning arrangements at the cafeteria were highly unsatisfactory. Raw, frozen chickens were delivered and allowed to defrost in cardboard boxes and plastic containers. After thawing they were spit-roasted, cooled, and put back into the cardboard boxes used for the raw chicken. After refrigeration, some cooked chickens were jointed and sold. For use in curries the flesh was removed from the chickens, cut into pieces and deep-frozen until needed. When needed, several portions would be taken from the deep-freeze and warmed by steam injection in stainless steel bowls. When warm, portions were removed as required, and served in curry sauce. The remaining portions were allowed to cool at room temperature, and rewarmed as required.

Salmonellae were isolated from chickens in the cardboard boxes and in the deep-freeze, from the stainless steel bowls and from trays in the shop. Wipe swabs from the cardboard boxes and cutting table also yielded salmonellae, and two of the five staff members tested were found to be excreting salmonellae.

The likely sequence of events was that the cooked chickens were contaminated from the cardboard boxes that had contained them when raw. Then the repeated steaming and cooling prior to consumption allowed the salmonellae to multiply. The infected food handlers were clearly infected by the food.

### *Faults*

- Cross-contamination allowed by placing cooked chicken in containers that had held raw food.
- Storing and handling chicken in changing temperatures allowing bacteria to multiply. Rewarming is particularly dangerous.

## Staphylococcal illness

### Example 1

A sandwich shop was involved in a series of complaints of staphylococcal illness. In the shop, large quantities of baked ham were sliced at one time, and the slices kept without refrigeration on the back bar of the serving area. The outbreaks were traced to the days when the owner of the establishment sliced the ham. Her hands were found to be contaminated with enormous numbers of staphylococci and her nasal secretions were also heavily contaminated.

### *Faults*

- Direct contamination of cooked food by staphylococci from a food handler's hands (or when sneezing, or blowing the nose).
- Storage of ham at room temperature for several hours, allowing bacterial multiplication.

### Example 2

A cafeteria was involved in seven cases of staphylococcal illness, resulting from contaminated corned beef sandwiches. Laboratory tests revealed that two employees who handled the corned beef had positive nasal cultures. One was the waitress who served the sandwiches. The other had a habit of coughing and putting his hand to his mouth while handling the corned beef.

### *Fault*

- Bad hygiene by the food handler allowed his hands, contaminated by staphylococci, to touch the food.

### Example 3

A barbecue was involved in a number of cases of staphylococcal foodborne illness. One of the food handlers had injured his hand by sticking a skewer into it while preparing pork. He was working wearing a hand bandage. Three days after his injury he had prepared and cut most of the cooked meat portions which were stored at ambient temperature until sold directly to the public. The food handler passed on staphylococci to the cooked meat from his injured hand.

### *Faults*

- Staphylococci were passed to food because the handler was not wearing a protective waterproof bandage.

- Cooked meat was stored at room temperature for several hours, allowing bacteria to multiply.

### Example 4

Of 110 people on a coach outing, 61 developed illness two hours after eating cold ham at a cafe. Staphylococci isolated from stools and vomit from some patients corresponded in type with those isolated from the ham slicer, the chef's hands, and other food in the cafe.

#### *Fault*

- Poor food hygiene standards. The ham slicer was not effectively cleaned, allowing proliferation of staphylococci in the cafe at room temperature.

## *Clostridium perfringens*

### Example 1

Abdominal pain and diarrhoea were reported in a large number of children in a school. The school kitchen was visited the day after the outbreak. The suspected meal, eaten 9 to 12 hours before the symptoms started, consisted of cold boiled salt beef, salad, and boiled potatoes, followed by pudding and jam.

The beef had been delivered to the kitchen on the previous afternoon in joints each weighing 1.8–2.7 kg. The meat was immediately cooked in large boilers for 2 hours and left in its own liquor all night to cool. The following day the meat was taken from the liquor, drained, sliced and served cold for lunch. A portion of this meat left over from the meal and kept in a refrigerator was examined in the laboratory. Several types of bacteria were found, including *Clostridium perfringens*.

#### *Fault*

- Leaving the meat in its own liquor all night at room temperature allowed the spores that survived the preliminary heat treatment to produce bacteria which multiplied vigorously.

### Example 2

Boiled chickens in liquor were transferred to open vessels, left all night to cool, and eaten cold. Illness followed the day after the chickens were eaten. *Clostridium perfringens* was found in a high proportion of dust samples from the kitchen. The organism may have been on the chickens before they were cooked or on the vessels where the cooked chickens were left to cool.

#### *Fault*

- Whatever the source of contamination, the long, slow cooling encouraged multiplication of *Clostridium perfringens*

### Example 3

Four large groups of people at a wedding party had dinners provided by a catering service. The total number of people served was 1100, and 320 became ill. Investigation revealed that the main item served was sliced roast beef and gravy. The gravy was prepared three days before the meals were to be served. The roast beef was reheated just before serving.

The gravy had been prepared in a large, single container, properly cooked and then placed, uncovered, in a refrigerator to cool. Steam from the gravy condensed on the bottom of pots and pans on the shelf above and dripped back into the gravy. The large amount of gravy in the container maintained its warmth for a long time, and provided an environment with a lack of air. This combination was ideal for the growth of *Clostridium perfringens*. On examination the gravy was found to be teeming with this organism.

#### *Fault*

- The gravy was held, uncovered, in a large container for an excessively long time (3 days) before the meal was consumed.

### Example 4

After eating a meal of roast beef and gravy, 150 people became ill, with severe diarrhoea and stomach pains. The beef and gravy had been prepared the day before, and allowed to cool in open trays without refrigeration for 22 hours. *Clostridium perfringens* organisms were found in both beef and gravy.

*Fault*

- Storage of beef and gravy in open trays at ambient temperature provided ideal conditions for the growth of *Clostridium perfringens.*

## *Clostridium botulinum*

### Example 1

An outbreak of botulism was reported in the owner of a restaurant, and two of his employees. The owner arrived at a hospital emergency room with symptoms of botulism. He died of pneumonia two days later.

The patient had eaten marinated fish prepared by his wife 15 days prior to his illness. She had stored the fish in three large, narrow-mouthed, glass jars with screw caps, and left them to cure under a table. When the investigators found the jars, a thick layer of oil had formed between the fish mixture and the air remaining in each jar.

The investigators attempted to find out who else might have eaten the fish and two more cases were identified. The second patient was a 24-year-old restaurant employee who had developed weakness six days before his employer, after eating portions of fish over a period of 3 days. He had shown symptoms of botulism but had recovered. Another employee, a 16-year-old boy, was found at home with severe weakness and some symptoms of botulism. He had also eaten small amounts of fish. Following therapy his condition improved slowly.

*Clostridium botulinum* type A toxin was found in serum from patients, and in the contents of the three jars of marinated fish.

The thick top layer of oil prevented the passage of oxygen to the fish. Anaerobic conditions, favourable to the multiplication of *Clostridium botulinum*, were created.

*Fault*

- Curing, bottling or canning of food that may contain botulinum organisms can be very hazardous. In this case, marination did not destroy the botulinum organisms present.

Apparently the pH of 4.6 was not low enough to inhibit spore germination and the growth of pathogens. The creation of an anaerobic condition in the jars allowed the formation of fatal botulinum toxin.

### Example 2

Nine people ate hamburgers in a restaurant in a small town. Seven of them ate sliced dill pickle on the hamburger, and two did not. All seven who ate the pickle became ill after returning home. Five were taken to hospital, and one died. Tests indicated *Clostridium botulinum* as the causative agent. The pickles had been canned by the restuarant proprietor.

#### *Fault*

- The temperature achieved in the cooking and canning process was not high enough to destroy the *Clostridium botulinum* organisms present.

## *Bacillus cereus*

### Example 1

An outbreak of foodborne illness occurred affecting eight people who had eaten in a restaurant. They all had a meal of soup, rice, prawns and bean shoots followed by ice cream, and were taken ill with vomiting 1½–2 hours later. *Bacillus cereus* was isolated from seven people who submitted specimens of faeces, the count in one instance being 2½ million *Bacillus cereus* cells per gram of faeces. None of the suspect rice was available but subsequent samples, prepared as usual, yielded over 30 million *Bacillus cereus* per gram.

### Example 2

A series of five small episodes of foodborne illness over a period of 2½ months affected customers eating meals in a restaurant. Thirteen people who ate fried rice with their meals became ill, whereas seven of their companions who did not eat the rice remained well. Illness was characterized by nausea and vomiting 1–6 hours after the meal, in all those affected, and diarrhoea after 2–5 hours in 8 people. *Bacillus cereus* was isolated in large numbers in stool specimens from the patients, in samples of fried rice and from boiled rice ready for frying. Left-over fried rice produced a count of 350 million *Bacillus cereus* per gram, and small numbers of the same organism were found in samples of uncooked rice.

#### *Fault*

- In some restaurants rice intended for frying is boiled the evening before it is needed, and allowed to dry off overnight at room temperature. This gives spores that survive the boiling process ideal conditions to germinate and multiply. The situation is made worse if new batches of boiled rice are added to the remains of old rice not fried on the previous day. Over a period of several day's enormous numbers of bacteria can be produced.

## *Vibrio parahaemolyticus*

### Example 1

Passengers arriving in London on an aircraft were found to be ill. Three passengers were immediately admitted to hospital and information about other cases was obtained. Three of the cabin crew who had left the aircraft at an intermediary stop were also ill. Everyone who was ill had eaten meals prepared in the town where the flight originated. Samples of complete meals prepared from the same batches of food served on the flight were frozen and flown to the United Kingdom for examination. *Vibrio parahaemolyticus* was isolated from cooked crab meat found in the hors d'oeuvre. The organism was of the same serotype as that isolated from stool specimens of the three patients in hospital. Raw meat from crab claws flown from the town of origin of the flight was also found to contain the same serotype of *Vibrio parahaemolyticus*.

#### *Faults*

- There are two possible faults. The organism may not have been killed by the cooking process. Alternatively, the cooked crab meat was recontaminated from raw crab meat during preparation of the dish. *Vibrio parahaemolyticus* is sensitive to heat, so contamination of the crab meat, after cooking, from sea water or uncooked marine products is the most likely cause of the illness.

## Typhoid fever

Over a period of two months, 72 cases of typhoid fever were reported to a local public health department. The average age of the patients was 19 years and an initial investigation revealed no common source of exposure.

All the patients had used water only from the municipal system and the city authorities reported no recent breaks in water lines for the area where most of them lived. A questionnaire was given to the first 25 patients in a search for common foods or food sources.

Analysis revealed four potential sources: ice cream cones bought from street vendors; food from two popular "fast food" establishments; and food from a specific tortilla molino (mill). At this mill, the two items purchased most commonly were corn tortillas and barbacoa (a Mexican barbecue dish).

Barbacoa is salted unspiced cow head cooked overnight under steam pressure. Meat from the cow head was deboned manually, by employees not wearing gloves, and held in a container on a heated grill at 71–79 °C.

Corn tortillas were prepared from corn kernels mixed with a lime slurry to remove cuticles. The mixture was boiled, washed and ground into masa (meal), which was then shaped by hand into tortillas. These were then heated for approximately 2 minutes on rollers warmed by gas jets and then sorted manually by employees.

A stool culture from one employee of the mill yielded *Salmonella typhi*. This employee worked in several locations in the mill including the area where barbacoa was deboned and the area where corn tortillas were shaped and handled.

### Fault

- Likely contamination from a typhoid carrier handling food after cooking and during final preparation.

## Shigellosis (bacillary dysentery)

### Example 1

Fifty-two cases of dysentery were reported among people eating in a dining hall. Shigella was isolated in a number of faecal specimens from infected people and from a cook. The cook had become ill with gastroenteritis about five days prior to the outbreak, but had continued to work.

### *Fault*

- The cook was an excreter and carrier of *Shigella* and may have contaminated working surfaces, utensils or food. He should not have been allowed to work during the illness.

## Viral gastroenteritis

Nausea, diarrhoea and fever were reported among people who had been to one or more of a series of eight receptions held over a week. Raw oysters, served at the receptions, were the suspected cause of the illness. Investigation revealed that, between them, 181 affected people had eaten about 950 oysters.

Examination of faeces revealed the presence of small, round viruses. A total of 4900 oysters had been transported direct from the fishery for the receptions. At the fishery the oysters had been purified by a depuration process as follows. The oysters were kept for 72 hours in a 5500-litre sea-water tank, where the water was continuously circulated through a 30-watt ultraviolet-light sterilization unit.

Routine bacteriological sampling of the oysters from the fishery over a period prior to the receptions had shown low levels of faecal bacteria. After the outbreak, however, a sea-water sample from the fishery area and a water sample from a river draining into the fishery area both revealed the presence of high levels of faecal bacteria.

### Fault

- Because of the presence of increased numbers of faecal bacteria in the sea-water, the depuration process (although successful in killing the bacteria in the oysters) was not fully effective in destroying the pathogenic viruses also present. Sufficient numbers of these survived to cause a gastroenteric illness when the raw oysters were eaten.

## Chemical poisoning

### Example 1

There was an outbreak of severe vomiting among people eating at a restaurant. Illness occurred a few minutes after eating. Investigation revealed that a badly worn, obsolete soda-fountain was being used. The soda-fountain allowed carbon dioxide to get into the fresh water system which was constructed of copper piping. The carbon dioxide caused copper from the pipes to be dissolved into the drinking-water. Soft drinks served at the restaurant contained enough copper to cause illness.

### Example 2

Customers in a large restaurant became ill during breakfast. Investigation revealed that the restaurant had bought a government surplus stock-pot which had been used by the night staff to mix reconstituted orange juice. The juice had been held in the pot for a number of hours before use. Laboratory tests showed that the pot was cadmium-plated. Cadmium from the plating had dissolved in the acidic juice.

### *Fault*

- Susceptible or worn metal piping and other worn metal surfaces had been allowed to come into contact with acidic liquids, causing the metal to dissolve and cause chemical poisoning.

Chapter 5

# Sources and transmission of food contaminants

Agents that cause disease (pathogens) can be transmitted to humans by a number of routes—air, water, direct person-to-person contact, and food. Some can be transmitted to food by animals, or by items of equipment. There are numerous possible routes of contamination and cross-contamination in food preparation areas. Cross-contamination is a very important concept in food safety. It occurs when contaminants are transferred from one food to another via a non-food surface, for example utensils, equipment, or human hands. Fig. 8 summarizes some of the most common routes of contamination, and Fig. 9 illustrates ways in which cross-contamination can occur, using poultry as an example.

Fig. 8. Common routes of cross-contamination.

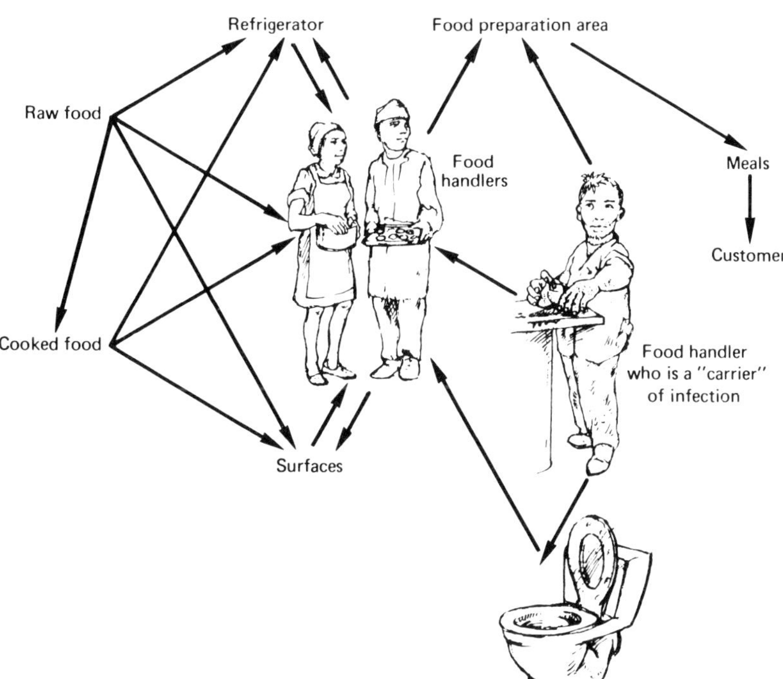

Fig. 9. Contamination risks with poultry.

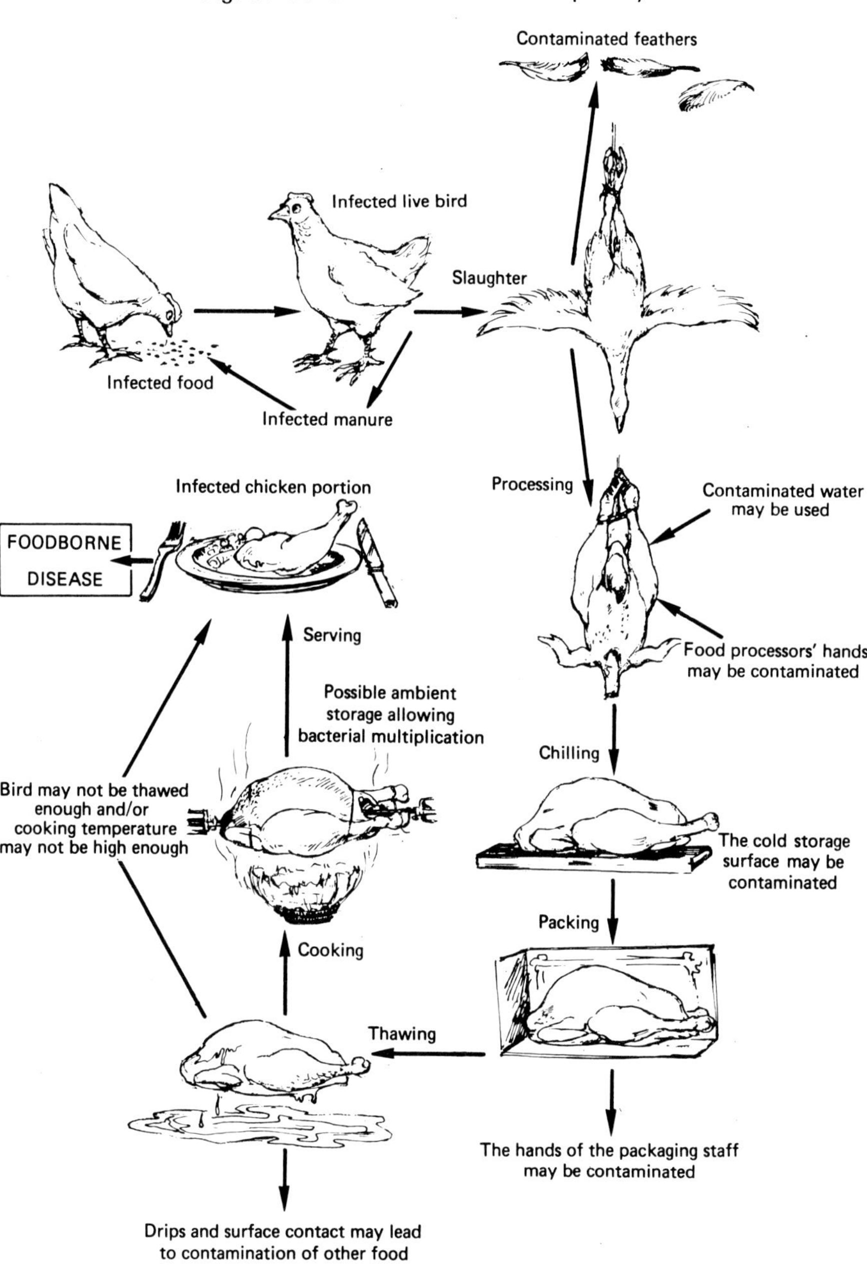

## Meat and poultry

Meat is a common source of pathogens. Pathogens in meat can be passed to people via the 'contamination chain' (outlined below and illustrated in Fig. 10).

An animal may have a subclinical infection and harbour pathogens in its tissues. These pathogens can be passed on to humans directly, if the animal is used for meat, or indirectly by passing the infection on to other animals through its faeces or body surfaces.

### The contamination chain

When an animal arrives at the slaughterhouse it will soil the pen where it is kept. Other animals using the same pen may pick up the contamination. When the animal is slaughtered, contamination may spread to the slaughterman's knife or clothing. Blood from the animal will contaminate the slaughterhouse floor and perhaps other working surfaces. The slaughterman may pass on this contamination to the carcasses of other animals. The carcass meat, when cut into joints, may contaminate knives, working surfaces, and workers' hands and clothing. The joint of meat, when it finally arrives in the kitchen, may drip tissue fluid on to prepared foods in the refrigerator or on work surfaces.

## Seafood

In many parts of the world, the sea receives much pollution, including human sewage, either directly from sea outfalls of sewerage systems or through rivers and drainage from land. Fish and shellfish, particularly molluscan shellfish, can easily pick up pathogenic organisms. Molluscan shellfish obtain their food from the sea by filtering sea-water through their bodies. In this way they trap pathogenic bacteria, which can easily be passed on if the meat of the shellfish is eaten raw, as it commonly is in the case of oysters and clams.

## Eggshells

Organisms, for example salmonellae, can be carried into food premises on the outside of eggshells. Food handlers can pick up infection while handling or breaking eggs. They should, therefore, wash their hands after handling eggshells. In some developed countries, there is increasing concern over evidence of the presence of salmonellae in eggs of contaminated birds.

Fig.10. The 'contamination chain' for meat.

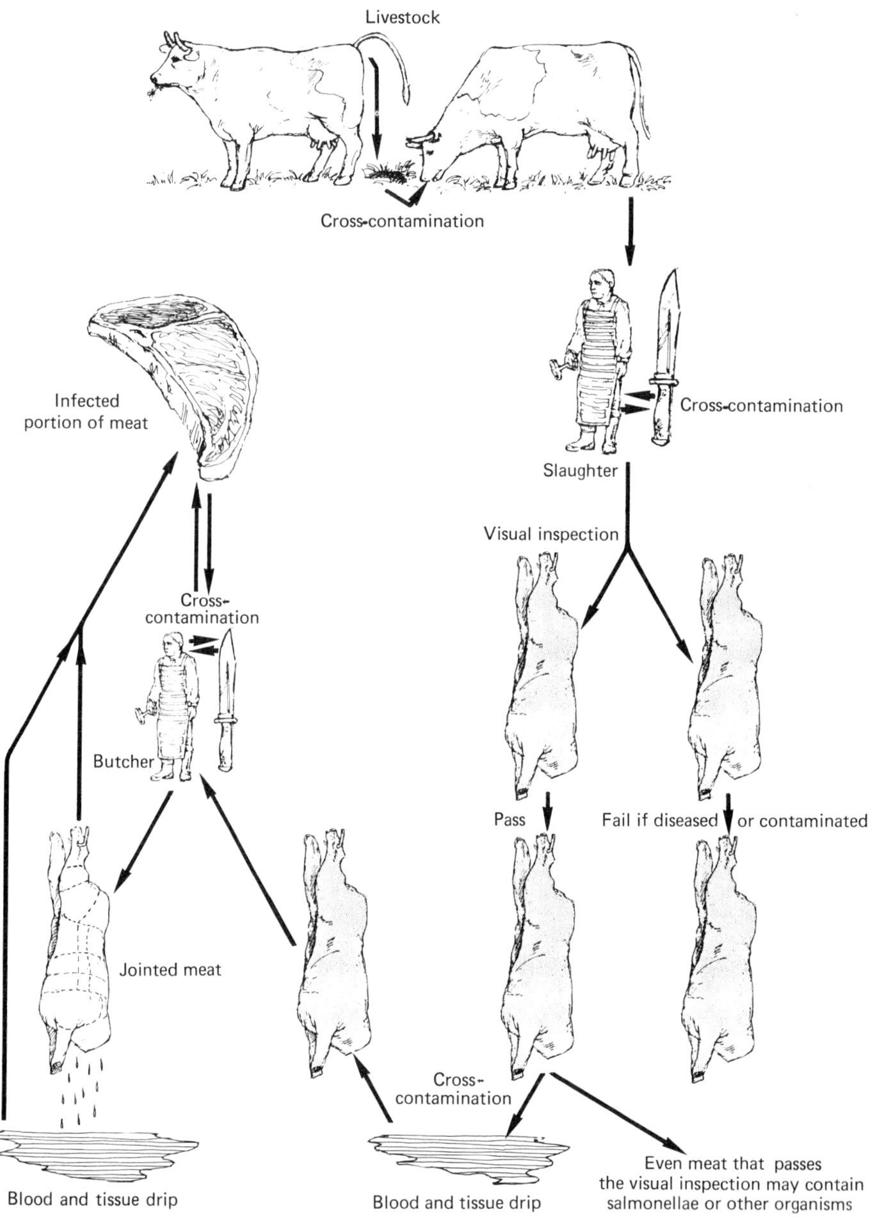

## Pets and other animals

The gastrointestinal tracts of animals may contain pathogenic bacteria. Salmonellae have been isolated from cattle, domestic pets, birds, terrapins, rodents, and many species of wildlife. Pets in domestic and commercial kitchens can easily pass on the contaminants in their bodies to food and food preparation surfaces. Rats and other rodents may be carriers of a number of diseases that may be transmitted to food through their urine and faeces, or through their saliva when they gnaw stored products.

## Insects

Flies and other insects also play a significant role in the transfer of infection. Flies are particularly dangerous because of their feeding habits. They mechanically transmit bacteria from one site to another by alighting on contaminated surfaces and then on food. Many other insect pests can cause similar contamination by direct contact between dirty surfaces and food, and some can also be of significance in food spoilage.

## Soil

*Clostridium botulinum* and *Clostridium perfringens* are found in the soil. They can be blown around by the wind, picked up by birds or animals, or taken up by plants and vegetables in their normal growing process. So anything grown in soil needs to be well washed before being taken into areas where food is prepared

## The human body

Infection can be passed from the human body to food. Staphylococci and other organisms may be found on the hands, under the fingernails, in cuts, burns, grazes, and in the nose and throat of infected people, and can easily be transferred to food, or food contact surfaces. Poor personal hygiene and lack of care in food preparation by infected persons can also transmit the hepatitis A virus.

## Animal feed

Feed for animals is made from the protein from animal carcasses, chicken feathers and a number of animal by-products, as well as other materials. It is therefore highly likely to be contaminated with pathogenic organisms of animal origin, either because of inadequate heat processing or subsequent recontamination.

Salmonellae can often be isolated from animal feedstuffs. The salmonella contamination originates in the animals, passes to the feed, which is then fed back to animals, in a recycling of contamination. It is likely that surfaces and people coming into contact with feed will also pick up salmonella contamination.

## Sources and transmission of food contaminants
## Important training points

- Cross-contamination risks require that all supplies of fresh meat, poultry and shellfish should be kept separate from cooked foods, on entering food premises and in final preparation and serving areas.
- Often cross-contamination cannot be avoided. As long as this fact is recognized, and food treated to eradicate the contamination, there is not a problem. Complacent food handling can however lead to cross-contamination becoming a danger.
- The hands of food handlers are often an important vehicle for transmission of contaminants. Attention to personal hygiene is essential.
- When shellfish are eaten raw, it is important that the sea and other water near the fishery are free of faecal pollution. Otherwise, there is a high risk of both viral and bacterial illness.
- Rodents, flies and domestic pets can contaminate both raw and cooked food at any stage of storage, distribution and final preparation.
- Animal feed is a common source of salmonella contamination in live animals. Cross-contamination in pens, and during slaughter will spread this contamination.

# Bibliography

BENENSON, A. S., ed. *Control of communicable diseases in man*, 14th ed. Washington, DC, American Public Health Association, 1985.

Botulism – Puerto Rico. *Morbidity and mortality weekly reports,* **27**: 356–357 (1978).

BRYAN, F. L. Factors that contribute to outbreaks of foodborne disease. *Journal of food protection,* **14**: 816–827 (1978).

GILL, O. N. ET AL. Epidemic of gastroenteritis caused by oysters contaminated with small round structured viruses. *British medical journal,* **287**: 1532–1534 (1983).

MOSSEL, D. A. *Microbiology of foods. Occurrence, prevention and monitoring of hazards and deterioration.* Utrecht, University of Utrecht, 1977.

NATIONAL RESTAURANT ASSOCIATION. *Food poisoning case histories.* Chicago, National Restaurant Association, 1964.

Restaurant outbreak of salmonellosis due to undercooked turkey. *Morbidity and mortality weekly reports,* **27**: 514–519 (1978).

SHEARD, J. B. Working party on food poisoning in hospitals, institutions and residential establishments. *Environmental health,* **93**: 263–266 (1985).

Typhoid fever, San Antonio, Texas. *Morbidity and mortality weekly reports,* **30**: 540–546 (1981).

WHO Technical Report Series, No. 598, 1976 (*Microbiological aspects of food hygiene*: report of a WHO Expert Committee, with the participation of FAO).

PART II

# PREVENTION OF FOOD CONTAMINATION

Chapter 6

# Structure and layout of the food premises

Clean and attractive premises of sound structure, and designed for ease of work, should promote food safety. Management need to understand the basic principles of good structure and layout so that, when necessary, they can play a constructive part in designing kitchens and work areas. Unfortunately many catering premises are not purpose-built or in the most suitable places, so careful planning is needed to make the most of what is available.

The amount of natural lighting and ventilation, access, possible storage areas, and the quality of the water supply are among the things that need to be considered when planning catering premises. For example, a separate area will be needed for storing refuse containers, and a pure and adequate water supply will be needed.

It is important that the air entering the premises should be clean, free from smoke and other pollutants, and that the surrounding area should not contain potential or actual breeding grounds for rats, mice, flies, or other harmful rodents or insects.

Premises should be planned so that they are easy to clean, and to keep clean. The further food has to be carried, and the more often it has to be handled, the greater the chance of it becoming contaminated. Ideally, premises should be arranged so that food can move in an orderly progression from the point of delivery to the areas used for preparation, cooking, serving and washing up (see Fig. 11).

A food delivery entrance, separate from the customers' entrance, is desirable. Ideally this entrance should open on to a yard, so that delivery vans can drive right up to the door. The yard should have an impervious, even, and well drained surface, a water standpipe, a hose for washing down the surfaces, raised and covered accommodation for refuse bins, bins for wet waste, and, if possible, a machine to compact packaging waste.

The vegetable store should be close to the food delivery entrance. It should be cool, dry, well ventilated, and large enough to allow for orderly storage. Vegetable preparation, whether carried out in a separate room or in part of the kitchen, should be done close to the vegetable store, yard and refuse bins, so that most of the dirt from the raw vegetables does not come any further into the premises.

There should also be a dry food store near the food delivery entrance which should be fly-proofed with removable 1.55-mm

Fig.11. Work flow diagram of a food service establishment.

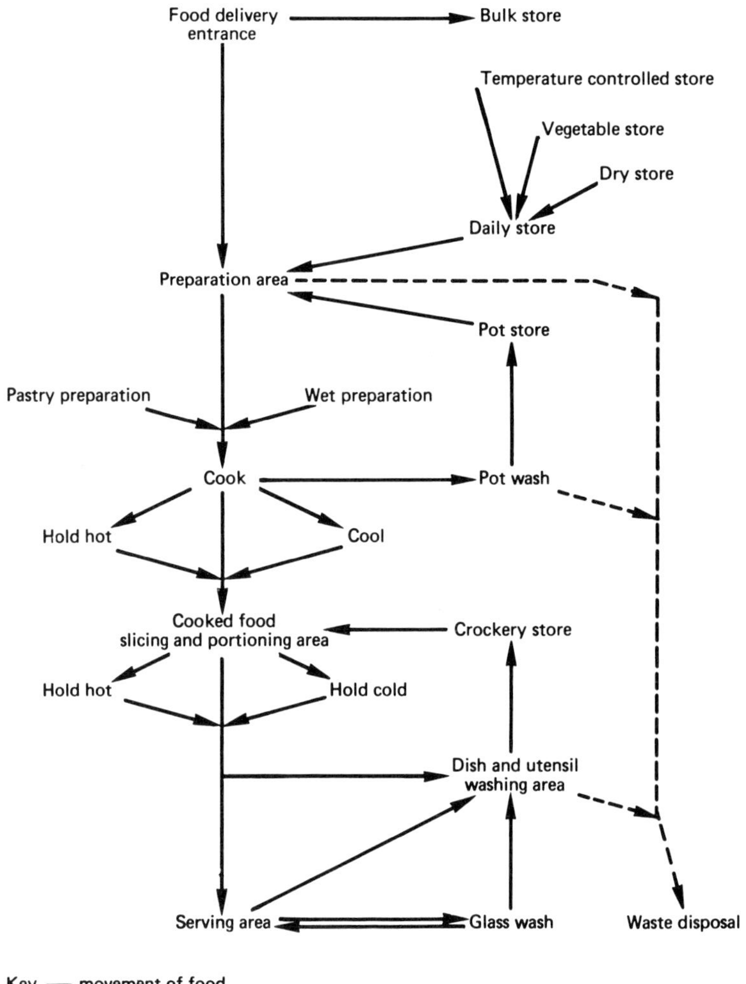


(16-mesh) screens over windows and doors. A temperature-controlled store for chilled meat and frozen foods is also useful near the food delivery entrance.

## Kitchens

### Structure

#### *Floors and drainage*

Floors should be built so that they are easy to clean and maintain. Damaged or uneven surfaces, and open joints, must be repaired promptly. Light alloy drainage covers are recommended because they are easy to remove for cleaning drains. Special attention should be paid to the cleaning of obstructed sites including areas behind ovens, boilers and other equipment, underneath tables and racks and at the junction of floors and walls. Any food spilt on the floor must be removed quickly to prevent staff from slipping. Drainage channels should be cleaned regularly, particularly ledges, grids and gulley corners.

#### *Walls*

Walls must have a smooth surface that can be easily cleaned. Textured paint finishes are not suitable. Wall surfaces in certain areas, for example around sinks and cooking equipment, must be resistant to excessive heat, moisture and physical damage. Well set, glazed ceramic tiles provide a good surface for walls. The upper part of the walls should preferably be cleaned every three months, and certainly every six months. The lower parts of the walls, particularly behind sinks and preparation surfaces, must be cleaned at least weekly, and more often if necessary.

#### *Ceilings*

Ceilings should be constructed of smooth, impervious material, and any infrastructure, for example beams, kept to a minimum. Particular attention should be paid to ventilation and lighting. Where ceilings are enclosed to cover pipes or ducts, access hatch ways or doors should be fitted so that regular inspection can be carried out.

#### *Doors*

Doors to all storage areas should be rodent-proof. Metal kick-plates should be fitted.

### *Windows*

Windows that can be opened should be fitted with fly-proof screens. Windows in roof spaces must be pest- and bird-proof.

### *Ventilation*

Adequate ventilation should be provided for cookers, hot plates, grills, etc. Any suspended screens, hoods, and ducts should be of materials and designs that are easily cleaned. Hoods and ducts should be inspected every three months or more often if necessary. Filters should be cleaned or renewed regularly because a build-up of grease may be a fire hazard. Ventilation systems, and the rate of inflow, should be checked annually and maintained in accordance with an agreed procedure.

### *Lighting*

All working surfaces must be well lit according to local regulations or standards. Attention should also be paid to the lighting in cold rooms and food storage areas.

All light fittings should be inspected every three months, or more often, if necessary. Regular cleaning is required, and could conveniently be carried out at the same time as the cleaning of the hoods and ducts of the ventilation system. The responsibility for maintenance and cleaning of light-fittings should be clearly allocated.

### Layout

The kitchen should never be used as a thoroughfare to other parts of the building. In planning, the chief factors to consider are the work-flow, the nature of the work, and the position of windows, doors and drains. It is important to make the fullest possible use of natural lighting and also the most efficient use of drainage.

Equipment should be placed so that there is plenty of room for cleaning. Narrow spaces, for example between cupboards or between equipment and a counter, are very difficult to clean and allow food debris to collect and attract insects.

Free-standing storage and kitchen units are much more hygienic than those fitted close to walls or in corners. An 'island' layout generally makes it easier to clean equipment.

Generally it is a good plan to have the work-tables against the walls between the sinks, and to have the ovens, stove and mixing machines in the centre of the room. Work tables should be movable for easy cleaning. Cooking-stoves and ranges usually need a canopy and an exhaust fan ventilation system to draw off fumes. Small extractor fans which draw steam and odours from small cooking ranges via a filter pad, and through activated carbon particles, can help ventilation and keep the working environment pleasant.

Where possible, areas where vegetables, raw meat or poultry are washed or prepared, and cooking units where steam is produced, should be located near an outside wall, to avoid long drainage channels passing through the kitchen. Ovens and stoves should stand on a solid foundation, preferably one with a concrete base and a surface that can be easily cleaned. In the dry preparation part of the kitchen there should be at least one deep, double sink with hot and cold running water. Automatic potato peelers should ideally be placed in a separate vegetable preparation area and should discharge their water into a drainage gulley. They should be fitted with an efficient strainer and trap for waste.

Pots and pans should be stored on racks or slatted shelves either upside down or on their sides. Other equipment should be stored in specified areas so that it does not clutter the kitchen.

### *Underground kitchens*

Underground kitchens present special difficulties. They are prone to flooding and drainage back-flow, and require special ventilation and lighting. Their windows should not open on to narrow areas or forecourts where dirt and rubbish could accumulate. This could lead to rubbish standing near food. It is also more difficult to prevent infestations in underground kitchens.

Rooms where food is stored need to be cool and dry. If an old building is to be used it is worth ensuring that the walls have an effective damp course.

## Working areas

Once again it is difficult to formulate a general rule about the amount of space required, but working areas should be large enough to allow employees to carry out their work comfortably, without feeling crowded at tables or having to queue for the use of sinks. The areas should not, however, be so large that extra walking is necessary. Employees will tend to neglect practices if they involve additional walking, waiting or working uncomfortably close to colleagues.

The kitchen, including the food preparation and washing-up rooms, should be at least half the size of the dining-room. Relatively more space than this is needed in very small establishments. A possible layout for the kitchen, food stores, food preparation areas and washing-up areas is given in Fig. 12.

## Washing-up rooms

Ideally there should be two washing-up rooms or areas—the pot-wash for heavy kitchen utensils that are likely to be greasy and an area for washing the glassware, crockery and cutlery used by the customers. Customers' glassware, crockery and utensils should not be washed in the pot-washing sinks. Both the pot-wash and the washing-up area should be at a distance from anywhere where food is being prepared or stored. Sinks used for washing utensils should not be used for the preparation of vegetables, meat or fish, or for hand-washing.

In small food premises it may not be possible to be as generous with the facilities for washing-up as outlined above. Extra care has to be taken in any circumstances where the washing-up sink is also used for food preparation and cleaning.

## Crockery store

This should be readily accessible from both the kitchen and dining-room. Plates, cups, saucers, dishes, basins, and other crockery should be stored in clean, dry cupboards or in a separate room, protected from dust, insects, and other sources of contamination. Plate racks should have drip trays underneath, connected to a drainage point. There should be reserves of glassware, crockery and cutlery to cover particularly busy times, and so that damaged items can be replaced promptly.

## Toilets and cloakrooms

Separate toilets should ideally be provided for employees and customers, although this may not be possible in very small establishments.

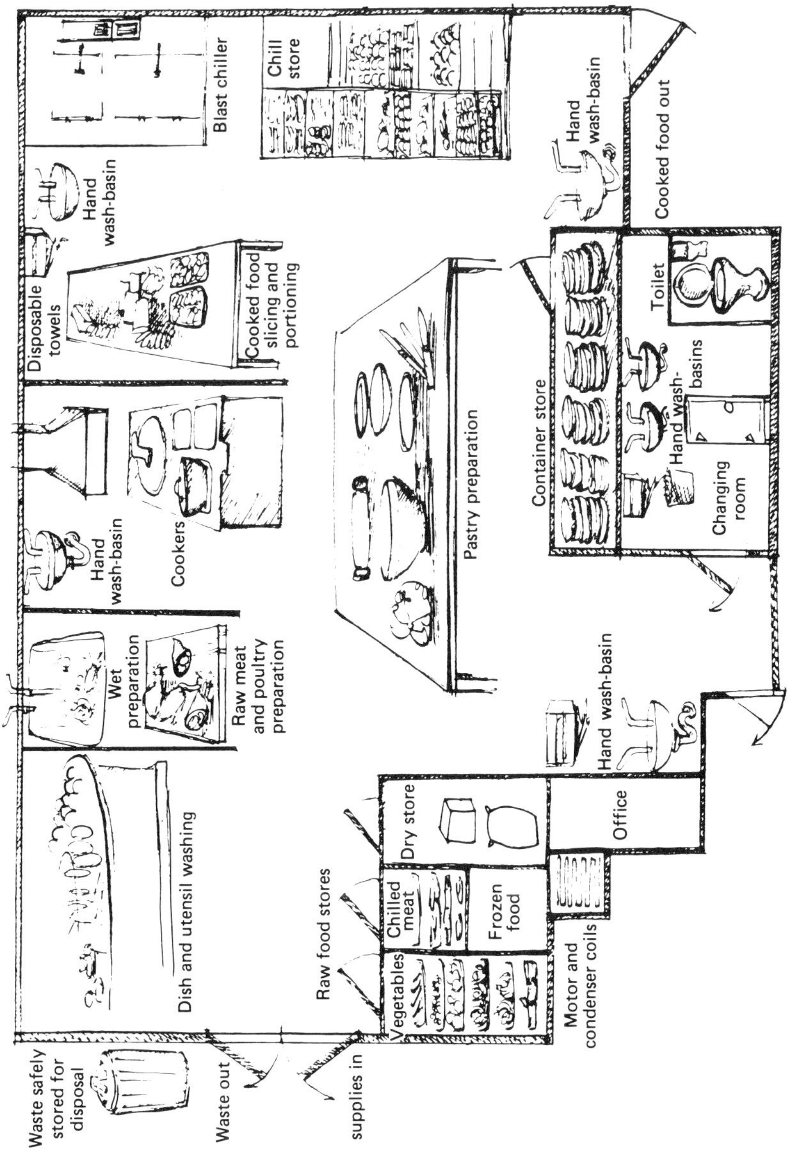

Fig.12. One possible layout for the food preparation areas of a food service establishment.

Toilets for catering workers should be readily accessible. Generally no employee should have to go more than 30 metres from the room where he or she is working to reach them.

The toilets should be well lit and separated from any workroom or dining room by an intervening ventilated space. Fully equipped hand wash-basins are required in or near the toilets, for example in the intervening ventilated space referred to above. Providing other wash-basins in working areas throughout the premises encourages staff to think about washing their hands frequently and makes it easy for them to do so.

Sinks where food is prepared or washing-up is done should not be used for washing hands after using the toilet. Notices should be put up on the premises making this clear.

A plentiful supply of warm water, soap, nail brushes and disposable towels should be available. Hot-air hand dryers should be well maintained. The use of special soaps containing bactericides can be useful in helping to promote the idea of clean and 'bacteria-free' hands.

Cloakroom or locker facilities are essential for the staff to keep any clothing and personal belongings not being worn during working hours.

## Dining-rooms

It is difficult to suggest precise standards of space needed for dining-rooms. This depends on the total number of meals served during a working period, the number of meals served at peak times when the dining-room is full, the type of meals provided, and the type of menu.

It is not desirable for customers to be crowded together in the dining-room. Waiters and waitresses need to have clear access to every part of each table so that dirty dishes and cutlery can be removed promptly, and so that the tables can be kept clean. Ideally dining-rooms should have enough space to accommodate racks for hats and coats, and shelves for parcels, books and gloves.

## Structure and layout of the food premises
## Important training points

- Kitchen storage areas should not be used for storing, for example, clothing, spare parts for machinery or old containers, that may harbour dirt, rodents and insects.
- The standards and cleanliness of toilets and washing facilities reflect the standards of the management and staff using them. Health or food inspectors are quick to notice this.
- Also, unlike most customers, health or food inspectors pay more attention to the cleanliness of the rear areas of food premises than to the dining-rooms. Yards, refuse storage areas and vegetable stores should appear clean and tidy, and should not contain long-standing waste materials.

Chapter 7

# Equipment

Equipment for food preparation should be kept in good condition, and be frequently cleaned and disinfected. Slicers and mincers should be cleaned thoroughly after each use. At the end of working periods, cream-making machines, soft ice-cream machines, and similar equipment should be cleaned in accordance with the manufacturer's instructions.

Raw meat and cooked meat must never be processed on the same machine without thorough cleaning in between. Utensils should not have wooden components or handles. They should be made of metal or other nonabsorbent materials.

## Crockery, cutlery, pots and pans

If these items are to be washed by hand, twin sinks should be used, one for washing and one for rinsing. This ensures efficient rinsing. Water containing detergent, and rinsing water, should be changed frequently. The rinse will have no disinfecting effect unless it is kept at 75–82 °C. This temperature is too high for bare hands, so baskets will be needed. Nylon brushes, washed and thoroughly dried between each use, are preferable to cloths. Cutlery should be machine-washed at a minimum temperature of 60 °C with a final rinse of 82 °C.

For other items low-temperature dishwashers are useful. They are compact and can operate on hot water from an ordinary supply system at a minimum temperature of 55 °C for both wash and rinse cycles. Three chemicals are used, detergent, rinse additive and disinfectant – usually sodium hypochlorite injected automatically in measured doses. The full wash cycle, involving a change of water each time, takes between $1\frac{1}{2}$ and 3 minutes. As well as reducing electricity costs the low-temperature dishwater is also said to improve the kitchen environment because the machine stays cooler and requires less extraction ducting. The machines are compact, fully automatic, electrically controlled, and can be installed very simply. Dishwashing machines need regular maintenance, and in some hard-water areas require water-softening equipment. Prompt attention is required if spray jets become blocked with hard-water deposits.

A system of heat-drying or air-drying in racks is recommended for crockery and cutlery. When items are dried by hand, disposable paper towels should be used. Damp plates should not be stacked for reuse as this delays drying and makes contamination more likely.

## Surfaces

Preparation surfaces should be impermeable, and constructed so that they can be cleaned thoroughly. Wooden surfaces should not be used. Cutting slabs and chopping blocks should be made of polypropylene or a similar material (see Fig. 13). The supporting structures must be maintained to a high standard, and should be regularly inspected and cleaned. Tables should stand away from the wall or be built into the wall. The junction of the table and the wall should be covered and sealed.

Food preparation and storage surfaces should be kept clean at all times. It is important that surfaces in direct contact with food are clean and dry before use, particularly if the food being prepared is to be eaten without further cooking.

Fig. 13. Preparation surfaces should be impermeable and easily cleaned.

British Airways

## Sinks

Sink units should preferably be made of stainless steel. They should be designed and sited so that any cleaning and maintenance needed behind or below the sink can be easily carried out. Ridges and joints at the junction of ledges and splash-backs (tiled areas behind sinks) should be avoided. It is recommended that areas behind sinks, draining-boards and adjoining walls should be protected by stainless steel sheeting, which is properly covered and sealed or jointed. Sinks, hand wash-basins and the areas above and below them should be well maintained and regularly cleaned.

## Refrigerators

Refrigeration facilities should be as far away as possible from stoves and other sources of heat. Refrigerators have thermal insulation to reduce the inflow of heat, but it is still advisable to install them in a cool place where the refrigeration mechanism will operate more economically. Refrigerators should not be boxed-in, as ventilation is essential for their mechanism.

Whatever the size, a kitchen requires proper refrigeration facilities for temperature-controlled storage of foods. The following refrigeration facilities are generally required.

- Cooling storage and hold-over store below 10 °C.[1]
- Chill storage spaces ranging from −3 °C to +5 °C.
- Deep-freeze storage at −18 °C and below.

If possible, there should be separate refrigerator storage for raw and cooked meats.

Refrigerators should be cleaned and defrosted regularly, and periodic inspections should be made for any defects requiring repair, or parts needing replacement. Temperature checks, using pocket and probe thermometers, should be carried out on foods stored in all refrigeration units, to ensure that they are being maintained at the required temperature.

## Vending machines

Vending machines for perishable foods, particularly made-up milk and meat dishes, should maintain an internal temperature of 7 °C

[1] Refrigeration at 10 °C will impede the growth of most pathogens. However, some (such as *Listeria*) will still grow, although rather slowly. For this reason, where possible, a temperature of 5 °C or below is generally advocated.

or less. The food should be packaged or placed in containers that reduce the risk of contamination. A thermometer should be placed so that it is easy to see without opening the machine. All foodstuffs stocked in vending machines should be properly rotated, i.e., the food that has been in the machine longest should be sold first. Perishable foods and made-up dishes should be discarded if unused after 24–72 hours, depending on the nature of the item.

All vending machines for drinks and perishable foodstuffs should be cleaned daily in accordance with the manufacturer's instructions. Their cleaning should be specifically allocated to a member of staff. Details of the cleaning instructions should be displayed next to the machine and should be strictly followed.

## Microwave ovens

Microwave ovens should be installed according to the manufacturer's instructions and should be checked at regular intervals to ensure that they are working properly, particularly that they cook or heat foods evenly, and that no excessive high frequency emission is occurring. Instructions for using these ovens, and cleaning instructions, should be prominently displayed next to the oven. Perishable foods and made-up dishes ready to be heated in microwave ovens should be stored under refrigeration at 10 °C or less. Although frozen foods can be defrosted in microwave ovens, they should preferably be thoroughly thawed first, and then heated in the microwave oven.

## Equipment
## Important training points

- Equipment (knives, slicers, chopping boards, etc.) used for cooked foods must never be used for raw foods, and vice versa, unless thoroughly cleaned between each operation.
- Remember that worn or damaged surfaces cannot be properly cleaned.
- To clean ventilation ducts effectively there must be access points. If none is installed initially, access doors or panels need to be fitted.
- If plates are rinsed at a high enough temperature (82 °C) and then stacked in plate racks, they will dry rapidly without further handling.
- When checking refrigerator temperatures, remember that the temperature of food in the cabinet may not be the same as the air temperature, particularly if the refrigerator has just been filled with food.
- Test microwave ovens on a variety of foods for heat distribution before using them. Make sure that the oven cooks or heats food evenly.

Chapter 8

# Maintaining a clean environment

## Heat

Heat is the most effective means of disinfection. Crockery, cutlery and equipment should always be washed in hot water. Dry heat for drying crockery etc. is also a good means of disinfection.

## Detergents and disinfectants

### Detergents

Detergents change the physical and chemical properties of water so that it can penetrate, dislodge and carry away residues that have hardened on food equipment. They reduce surface tension and are good foaming, wetting and emulsifying agents.

The choice of detergent will depend upon: the substance to be removed; what material the article or surface to be cleaned is made of; whether hands will come into contact with the solution; whether it will be used in a machine; and the chemical nature of the water, such as its degree of hardness. Most detergents are cleaning agents only, with little or no bactericidal properties. However, some combine a cleaning substance with hypochlorite so that cleansing and disinfection can both be accomplished by one liquid or powder. These mixtures are useful in small establishments with limited washing-up facilities. They may also be used by food vendors for washing hands.

#### *Synthetic detergents*

General-purpose synthetic detergents are mildly alkaline and are effective for removing dirt from floors, walls, ceilings, equipment and utensils. Heavy-duty detergents are highly alkaline and used for removing wax and baked-on grease. Detergents used in dish-washing machines are also highly alkaline, but those intended for hand-washing are neutral, and contain ingredients to make them mild for the hands. Certain types of dirt are not affected by alkaline cleaners, for example lime encrustations in dishwashing machines, and rust stain and tarnish on copper and brass. Acid cleaners, sometimes in formulas also containing detergents, may be used in these cases. These must be selected and applied carefully to avoid damage to the surface being cleaned or the skin of the user.

### *Soap*

This is a simple detergent, generally used for personal washing. It lacks the good wetting action of synthetic detergents and the powerful dissolving properties of alkalis. In hard water it may form a scum, and lather with difficulty. The foam produced breaks up easily.

### *Abrasive cleaners*

When oil is attached so firmly to a surface that alkaline or acidic cleaners will not work, a cleaner containing a scouring agent (usually finely ground feldspar or silica) may be used. Badly soiled floors, and worn and pitted porcelain, benefit from the use of these cleaners. They must be used with care on smooth food-contact surfaces, as they may cause damage.

## Choosing a detergent

In whatever way it is used any detergent must be able to:

- Throughly wet the surface to be cleaned.
- Remove the dirt from the surface.
- Hold the removed dirt in suspension.
- Be easily rinsed.

Additional questions that may need to be answered before selecting a detergent are:

- What are its bactericidal properties, if it is combined with a disinfectant?
- Is it corrosive?
- Does it prevent scale formation?
- Is it economical?

However, it is unlikely that any single detergent will meet all the requirements.

## Disinfectants

Disinfection reduces the number of living microorganisms. It does not usually kill bacterial spores. No disinfection procedure can be completely effective unless it is preceded by thorough cleaning.

### *Hypochlorites*

These are good disinfectants for use in food premises; they are inexpensive and leave little taste or smell if used at the correct concentration. They have a wide range of antibacterial activity, including activity against bacterial spores, a property not exhibited by most other disinfectants. However, they can be inactivated by some organic materials. Strong solutions can corrode some metals, especially aluminium alloys.

### *Iodophor disinfectants*

These incorporate iodine with a detergent. They tend to be inactivated by organic material. They are less effective against spores than are hypochlorites, and more expensive, but leave little taste or smell.

### *Quaternary ammonium compounds (QACs)*

These disinfectants are less effective against bacteria than hypochlorites or iodophors. Fresh solutions of disinfectant should be prepared daily in clean heat-treated containers.

### *Amphoteric surfactants*

These disinfectants have detergent and bactericidal properties. They are of low toxicity, are relatively noncorrosive, tasteless, and odourless. They may, however, be inactivated by organic matter.

### *Phenolic disinfectants*

There are several types. White, fluid phenolics and clear, soluble phenolics have a wide range of antibacterial activity similar to hypochlorites and iodophors. They are not easily inactivated by organic materials, but may be inactivated by plastics and rubber. Some brands have a powerful smell and may leave a taste on food. They are not generally recommended for use inside food premises.

## Choosing a disinfectant

- Choose heat for disinfection wherever possible.
- Only use a chemical disinfectant when it is impossible to apply heat.

- Arrange for equipment and surfaces to be cleaned before heat or chemical disinfection.
- When a chemical disinfectant has to be used, choose one with a wide range of antibacterial activity. Hypochlorite is good for general use. Normally dilutions containing 100–200 mg of available chlorine per litre are adequate. In cases where absolute cleanliness cannot be assured a dilution containing 1000 mg/l or more is recommended.
- Dilutions of chemical disinfectants should be made fresh daily, or when required, in clean, dry, heat-treated containers.

### Handling and use of detergents and disinfectants

Care should be taken when handling detergents and disinfectants. The manufacturer's instructions on handling and dilution of fluids must be strictly followed at all times. Under no circumstances should food handlers attempt to formulate their own detergents. Acid cleaners must be selected and applied carefully, to avoid damage to the surface being cleaned, or the skin and clothes of the user.

A purpose-built cleaner's room or storage area of adequate size should be provided for the storage of all cleaning equipment, including floor scrubbers, laundry bins, supplies of cleaning materials, detergents, soaps, and disinfectants. Depending on the size of the premises, the room should be equipped with a sluice tank, hot and cold water supplies, and a rack for drying mops.

## Refuse disposal

Wet refuse should be disposed of, when possible, in a sink disposal unit connected to the drainage system. Where such a unit is not practical or if large quantities of wet refuse are involved, an adequate system for the temporary storage of refuse in the kitchen should be provided. It is recommended that plastic or paper bags are used. Paper sacks should be of good quality, and made from impregnated paper to give adequate strength when wet. Plastic bags should be of heavy gauge polythene. To reduce handling, stands and holders with close-fitting lids should be provided for refuse sacks.

Where possible, waste-disposal units of suitable size and capacity should be provided in the kitchen and in central dishwashing areas. Waste food should be returned to the central wash-up area after each meal, and finally disposed of through the waste-disposal unit. If no unit is installed, waste should be collected in plastic bags, which should then be sealed. These sealed bags should be removed at once and kept in storage containers. These large bins or containers, in which waste or refuse is to be carried away from the food premises, should be kept out of food preparation or kitchen areas, in a special refuse storage area. This area should be purpose-built and well-ventilated. Walls should be of good quality materials, and rendered to give a smooth nonabsorbent finish. The area should be paved and roofed, and provided with adequate drainage so that any paper or sacks stay dry. A water supply point should be available for cleaning the area.

Refuse should not generally be burned near food premises. However, if there is no other means of disposal, a smokeless incinerator should be used. A refuse compactor is ideal for reducing the space taken up by plastic and cardboard packing materials. Before installing a compactor it is important to ensure that the authority or agency responsible for collecting refuse can cope with heavy loads of compacted material.

## Pest control

Proper measures to protect hotels and restaurant kitchens from rodents and insects can considerably reduce the risks of food becoming contaminated and being wasted. Measures will fail, however, if employees neglect to shut doors or replace the lids on wastebins.

### Rats and mice

Rats and mice are destructive, and are dangerous sources of infection. They breed rapidly, destroy food in fields and stores, and carry and transmit pathogenic bacteria. Any surface they touch must be regarded as contaminated. Constant and careful watch must be kept for signs of infestation. Small numbers of rats and mice may be difficult to detect. All staff, particularly cleaning staff, should be instructed to report immediately any signs that they may notice in the course of their normal work. Useful signs are droppings, smears and runs, holes and scrapes, gnaw marks, grease marks on skirting boards, footprints in dust or moist earth, damaged food or food containers, and live or dead rats or mice.

It should be the duty of management regularly and systematically to examine any part of the premises that is not generally in view, for example warm, dark corners; passages; stairs; under-stairs cupboards; the shafts of service lifts; floor spaces; spaces beneath shelving; areas behind piles of stock in the food store; lofts and crevices; and openings in walls and ceilings where pipes pass through. Outbuildings, waste ground, and other possible breeding sites should also be observed.

To stop rats and mice entering the premises, the building must be kept in sound repair and all holes, drains, and other possible points of entry sealed (see Fig. 14). Wood panelling, false ceilings, and boxed-in pipe work in food preparation rooms provide shelter for rats and mice and lead to infestations. Places gnawed by rodents trying to get into the premises should be covered with metal.

Rodents may enter premises in sacks of food such as flour, in straw packing materials, in cartons and boxes, or on vehicles. Faulty methods of food storage, careless stacking, and general untidiness may then allow a considerable infestation to become established in a sealed building. Materials stored in the open, in sheds, or in outbuildings, should be stacked on wooden pallets at least 30 cm above the ground or floor, and 60 cm from the wall.

Rats drink about three times the amount they eat; so denying them sources of water, such as dripping taps and defective gutters, is important. Material that is likely to be suitable for rat food, for example cereals, starchy vegetables, and fatty compounds, including tallow and soap, should be kept in metal, rodent-proof bins or containers.

Refuse awaiting removal should be stored in properly covered metal dustbins. Cartons, boxed goods, and sacks should be neatly stacked close together on wooden pallets, at least 30 cm from the floor and 60 cm from the wall.

Empty food containers should be throughly cleaned before they are put aside for collection or reuse. Food scraps, crumbs and refuse, such as peelings, cores, and husks, should be swept up at the close of business each day and placed in rubbish bins. Areas outside the premises should be kept clean and free of material that might form a breeding ground for rodents.

When premises are infested, steps should be taken immediately to destroy rats and mice. There are two effective means of doing

Fig. 14. Broken drains and drain grids can lead to rodents gaining entry to food premises.

Rentokil Ltd

this—poisoned baits and traps. Domestic pets such as cats should not be used as a deterrent against rodent infestation in food premises, as they themselves may cause foodborne illness. Traps are not effective against large infestations, although they may be useful in eliminating the survivors of a poisoning treatment, or to stop a building being reinfested after it has been cleared. Pesticide dusts may also be blown into holes and cavities used by rodents, and fumigation with gas may be necessary on ships or in warehouses.

Rodent poisons are dangerous to human beings. All rodent treatments should be carried out by experienced, qualified personnel employed by a local health agency or by a specialist firm. Anticoagulant poisons are best because of their low level of toxicity for people and domestic animals. They include diphenadione, pindolol and warfarin. Anticoagulant-poisoned bait must be available to the rats for at least two weeks to obtain satisfactory results.

## Flies and other insects

A wide variety of insect pests, including houseflies, bluebottles, greenbottles, cockroaches, ants, wasps, and mites, are attracted to food premises. Some of these insects are significant in the spoilage of stored food.

Control measures to prevent insects infesting food premises usually involve the following:

- Protecting the building against entry.
- Eliminating breeding places.
- Protecting food so that there is no risk of access to food by insects.
- Destroying the insects at some period of their life cycle, either outside or inside food premises.

Flies, bluebottles, cockroaches, and ants can carry germs from refuse or excrement to food and cause foodborne diseases. Flies carry bacteria on their bodies and legs. In feeding they can contaminate food by regurgitating partly digested food back on to food surfaces.

Insects may drop into food, and meat can become contaminated with insect eggs or larvae, leading to complaints from diners and perhaps action being taken against the food service establishment for the sale of dirty food.

It is virtually impossible to stop insect pests entering buildings although wire mesh screens on open windows, doors, and ventilators, will keep out most flying insects. Such screens should be 1.5 mm (16 mesh), and made of copper, brass, galvanized steel or nylon. All mesh screen doors should open outwards, and staff should not open or remove them to obtain more ventilation. As with rodents, breeding sites need to be eliminated.

Insect pests are normally destroyed by the use of insecticide powder or sprays against the eggs, larvae or adults. Where their use is permitted, sprays that leave a lasting residue (residual sprays) will control most species of flies. However, care must be taken with their use in food preparation areas. They should be kept away from all uncovered food.

Vegetation, walls, refuse containers, and any areas near the food premises, may be sprayed with residual insecticides to control adult flies and larvae. Pyrethrum may be used as an indoor spray for quick knock down of flies. The effect of synthetic pyrethroids lasts longer when applied to surfaces on which flies alight.

Devices using fluorescent tubes to attract flying insects on to an electrified grid are useful. The insects are killed immediately and fall into a collecting tray, eliminating the danger of dead insects falling into food.

Cockroaches live and breed in moist, dark places, for example around plumbing (see Fig. 15), under refrigerators, and in cupboards and pantries. Outdoors they are found in piles of debris. Cleanliness is the first measure in eliminating cockroaches. All food handling areas should be cleaned frequently.

Residual insecticides can be sprayed into places where cockroaches cross or hide. Painting a band of insecticide 10 cm wide at the bottom of walls where they join the floor has also been found to be effective. The band must be continuous to ensure contact with the cockroaches as they enter or leave the room. Boric acid powder can be blown into cavities harbouring cockroaches.

Fig. 15. A typical breeding-ground for cockroaches.

Rentokil Ltd

Aerosol formulations of pyrethrum can be useful to flush insects from their hiding places and locate sites of infestation. Longer-lasting insecticides can then be applied to these hiding places. Special care needs to be taken to prevent food or food contact surfaces from becoming contaminated with pesticides.

## Pest control contracts

Managers need to consider certain safeguards when entering into a pest control contract with a private contracting company. Knowledge of poisons and insecticides, the habits of various pests, and the risks of food contamination is essential. Before selecting a company the following aspects should be considered.

- Is a survey to be carried out before treatment, and will the survey report be submitted to the management and discussed before treatment starts?
- What type of pests will the treatment cover?
- What advice will be given on preventive measures?
- How quickly can treatment begin, and how frequently will visits be made when infestations occur?
- What staff and equipment does the contractor have?
- How frequently will routine visits be made, and what will they involve?
- Will reports be made on routine and follow-up visits?
- Can the contractor provide the names of three current customers?
- Will the methods and materials to be employed be of an approved type?
- What training have the staff been given?
- What public liability cover does the contractor have?
- Is the contractor a member of a pest control association that observes a code of practice?
- If the contractor finds pests not covered by the contract, will management be notified?
- Will the contract include servicing of mechanical or electrical pest control equipment?

Before entering into the contract, management should agree on a contact person who will discuss operational aspects with the contractor and receive reports on routine visits.

## Maintaining a clean environment
## Important training points

- Hot water, combined with a moderate amount of detergent, removes grease from crockery and cutlery.
- A detergent emulsifies and loosens fat from articles being washed.
- Disinfectants are not always necessary, either for general cleaning or for dishwashing. Effective thorough cleaning is more important.
- Rats, mice and insects are particularly attracted to premises where bad housekeeping and untidy habits are common.
- It is just as important to spend money regularly on maintaining buildings to keep rodents out, as it is to employ contractors on a regular basis to carry out disinfestation treatment.
- It is of paramount importance that chemicals and other poisons used for the control of pests do not contaminate food. Any chemicals stored in the food premises should be in sealed lockers to which only designated people have access.
- Staff should not try to eliminate rats and mice by the use of traps or poisoning; this must be left to experts. However, members of staff must be trained to recognize an infestation so that the pest control experts can be alerted.

Chapter 9

# Personnel

A major risk of food contamination lies with the food handler. Dangerous organisms present in, or on, the food handler's body can multiply to an infective dose, given the right conditions, and come into contact with food, or surfaces used to prepare food.

There has been some speculation as to whether most cases of foodborne disease arise from contamination from a carrier employed as a food handler, or whether the food handler acquires the contamination from the food that he or she handles. In fact, in virtually all cases involving infected food handlers, they have acquired the infection from food handled in the course of their work.

Nevertheless, food handlers infected or colonized with pathogens may contaminate food, thus transmitting foodborne illness. To minimize this risk, cost-effective measures, for example the education and training of food handlers in personal hygiene and safe food handling techniques, should be instigated.

## Health surveillance

It has been concluded that routine medical and laboratory screening of food handlers is of no value, as such examinations only reveal the health status of the worker at the time of examination and cannot take into account subsequent bouts of diarrhoea or other infectious conditions. Medical examinations are also unreliable in the detection of carriers of pathogens who, in most cases (with the possible exception of those excreting *Salmonella typhi*), are unlikely to transmit gastrointestinal organisms. Food handlers should, however, be encouraged to report immediately if they are ill.

There is no evidence that food handlers infected with human immunodeficiency virus (HIV) transmit the virus through food. Therefore, routine serological testing for HIV is not relevant. A commonsense approach to monitoring the health of staff can be taken, in any establishment, by following the recommendations below. Health monitoring needs to cover all staff on the premises. Special arrangements may be needed to cover contract staff who visit food handling areas, for example to clean or service food vending machines.

## Health questionnaires

A questionnaire, similar to the one in Fig. 16, should be completed by all job applicants, and then looked at by a doctor

Fig 16. Sample health questionnaire for prospective food handlers.[1]

**HEALTH QUESTIONNAIRE**
**CONFIDENTIAL**

Please answer the questions below as fully as possible and return the form to Personnel.

Please tick yes or no. If you have received treatment please give the date and the name and address of the hospital or doctor.

| | | |
|---|---|---|
| Have you ever had enteric fever (typhoid or paratyphoid) | No ☐ | |
| | Yes ☐ ________ | Date ______ |
| | ________ | |
| | ________ | |
| Have you had diarrhoea and/or vomiting for more than one day within the last 7 days? | No ☐ | |
| | Yes ☐ ________ | Date ______ |
| | ________ | |
| | ________ | |
| Are you suffering from any of the following: | | |
| Skin rash? | No ☐ | |
| | Yes ☐ ________ | Date ______ |
| Boils? | No ☐ | |
| | Yes ☐ ________ | Date ______ |
| Discharge from the eye? | No ☐ | |
| | Yes ☐ ________ | Date ______ |
| Discharge from the ear? | No ☐ | |
| | Yes ☐ ________ | Date ______ |
| Discharge from the nose? | No ☐ | |
| | Yes ☐ ________ | Date ______ |
| Have you ever lived abroad? | No ☐ | |
| | Yes ☐ ________ | Date ______ |
| | ________ | |
| | ________ | |
| Have you travelled abroad in the last 3 weeks? | No ☐ | |
| | Yes ☐ ________ | Date ______ |
| | ________ | |
| | ________ | |
| Please give the name and address of your general practitioner (personal doctor) | Dr ________________ | |
| | ________________ | |
| | ________________ | |

Your name and address:

________________ Date ________________

________________ Signature ________________

________________

________________

[1] Adapted from: LONDON BOROUGH OF HOUNSLOW. *Health monitoring arrangements for food handlers*. Environmental Health Codes of Practice, No.1.

acting on behalf of the food service establishment. After reading the questionnaire, the doctor can assess whether further examination, or treatment, is necessary. It may be decided, because of the medical history, that the person should not be employed at all in food handling.

## Food handlers' agreement

All food handlers should agree to report to their employers any infection or circumstance that might lead to food contamination. An agreement, to this effect, can be drawn up for food handlers to sign. A suggested model agreement is given in Fig. 17.

### Doctors' card

It may be necessary to alert a food handler's doctor to the fact that his or her patient is employed in the preparation of food, because an illness in the food handler may lead to food contamination. A card can be issued to staff, for presentation to their doctor at each visit, stating that the patient is a food handler; it can also contain a reminder to the food handler of his or her obligation to report infection. A suggested wording for such a card is given in Fig. 18.

### Contingency plan in case of outbreaks

A contingency plan should exist in case there is an outbreak of infection among the staff in a food service establishment. The plan should make clear who is responsible for the following actions:

- Identifying all staff affected and excluding them from the premises until they are clear of infection.
- Identifying, withdrawing and retaining any suspect food for examination.
- Communicating with the local public health authority, who will set up any investigation necessary following such an incident.

## Personal hygiene

Food handlers should be encouraged to adopt a code of personal hygiene that could be outlined in a leaflet or booklet to

Fig. 17. Model agreement for food handlers to report illness.[1]

Please read this agreement carefully, sign it, and return it to your manager or supervisor.

I agree to report to the employer if I suffer an illness involving any of the following.

Infectious hepatitis (viral hepatitis A)
Diarrhoea
Vomiting
Fever
Sore throat
Skin rash
Other skin lesion (boils, cuts, etc., however small)
Discharge from ear, eye or nose.

I agree to report to the employer before commencing work if I have suffered from any of the above conditions while on holiday.

I accept that I may be required to inform my employer when I return from travelling abroad.

I have read (or had explained to me) the rules on personal hygiene and hygienic food handling practices.[2]

I understand that failure to comply with this agreement could lead to disciplinary action.

Signature. . . . . . . . . . . . . . . . . . . . . . . . . . . . . . . . . . . . Date . . . . . .

[1] Adapted from: LONDON BOROUGH OF HOUNSLOW. *Health monitoring arrangements for food handlers*. Environmental Health Codes of Practice, No.1.

[2] Leaflets and instructions on personal hygiene should be given to all workers when they are first employed.

be given to staff when they take up employment. A suitable booklet may be obtainable from the local health authority, or managers could produce their own using the following, or similar, wording.

- Wash your hands frequently, particularly at the following times:
  after visiting the toilet for any reason,
  immediately before handling food,
  immediately after handling refuse,
  when your hands look dirty,

Fig. 18. Sample doctors' card for use by food handlers.

To the employee

You are required to inform your employer if you have any of these conditions, whether you visit the doctor or not.

Infectious hepatitis (viral hepatitis A)
Diarrhoea
Vomiting
Fever
Sore throat
Skin rash
Other skin lesions (boils, cuts, etc., however small)
Discharge from ear, eye or nose.

Also please notify your employer if you are going abroad, or you are in contact with infectious disease in your immediate family, friends, or workmates.

Please fill in your name, work address and telephone number on the reverse of this card, and ask your supervisor or manager to sign it in the space provided.

**Show the card to your doctor whenever you make a visit.**

To the doctor

The bearer of this card is a food handler at:

______________________________
______________________________
______________________________

Tel: _______________

You will be aware of our responsibilities regarding infectious conditions that could be transmitted via food, and I trust this information will be helpful when you decide on any treatment necessary.

Signature ______________________________
(of manager/supervisor)

when you know your hands are contaminated,
immediately after handling pet birds or animals.

- Keep your clothes and overalls clean, and wear the protective clothing provided.
- Always keep your hair covered so that stray hair or dandruff cannot get into food. Do not comb or tidy your hair in the food preparation area.
- Avoid wearing rings and bracelets when you are handling food. Wedding rings can be an exception.
- Cover all cuts and grazes completely with waterproof dressings.
- Try not to smoke. If you must, always leave the food preparation area, and wash your hands before returning.
- If you have one of the following conditions report it to management:

  infectious hepatitis (viral hepatitis A)
  diarrhoea
  vomiting
  fever
  sore throat
  skin rash or other skin lesion, for example boils or cuts
  discharge from the ear, eye or nose.

- If you are handling food that is ready to eat, use tongs and not your fingers.
- Do not cough or sneeze over unscreened food.
- Pick up knives and forks by their handles, glasses by their stems, and plates by their edges.
- Clean up work areas as you go.

The main points from the booklet could be summarized on charts and displayed in the workplace (see Fig. 19 and 20).

Daily hand inspections to look for incorrectly dressed or infected hand lesions should be encouraged for staff who handle cooked or ready-to-eat food. No member of staff who has an infected lesion of the hand, or the arm below the elbow, should be allowed to handle food until the lesion is completely healed.

## Protective clothing

Management should set an example by always wearing protective clothing in food handling areas, as should all visitors to the premises.

Fig.19. A teaching chart on personal hygiene that could be displayed in staffrooms or inside clothing lockers.

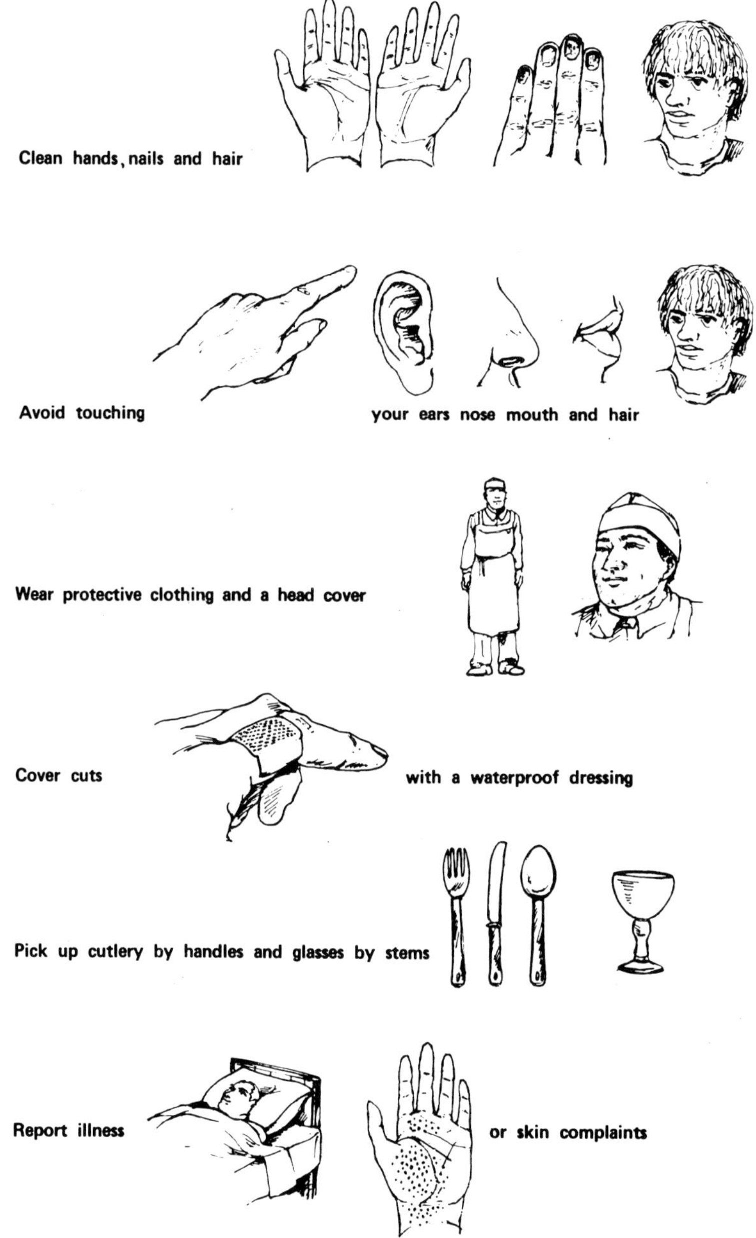

Fig.20. Illustration from hygiene instruction booklet that could be displayed in kitchens, work areas or toilets.

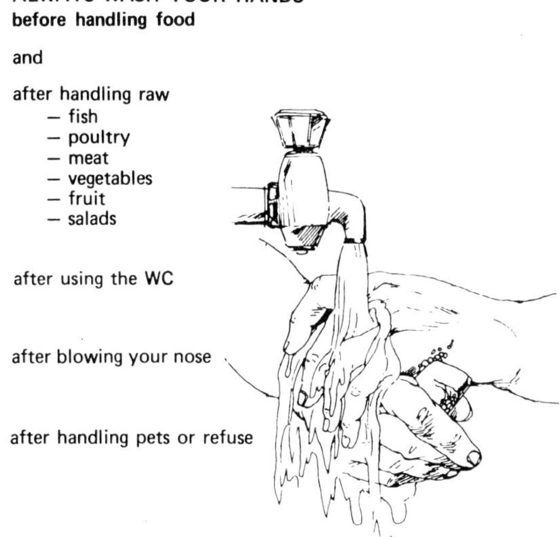


Clothing should be light in colour, changed frequently, and made of material that can be easily washed and kept clean. Uniforms or coats of drip-dry fabric ease the work of daily laundering and it may be possible to hire them on a daily or weekly basis from specialist laundry companies. Protective clothing should adequately protect both the food and the wearer.

Serving staff should wear light-coloured, washable dresses or overalls. Nylon clothing has the advantage that it can be washed at the end of the working day, dried overnight, and needs no ironing. However, it may be unsuitable in hot climates. Caps and other head covering should effectively protect food from contamination by hair.

Cooks, and those engaged in preparing and serving food, should have white or light-coloured overalls that ensure that food will not come into contact with any clothes worn underneath. It is customary for cooks to wear white caps to protect the food from hair as well as to protect the hair and scalp from the effects of steamy heat, fat vapours and flour. For washing up and for vegetable preparation, such as potato peeling, rubber aprons are a good idea. Personnel who have to stand for long periods, should be given the opportunity to learn how to stand

and walk with the minimum amount of fatigue and discomfort. They should wear comfortable, strong shoes or boots.

- Shoes or boots should support the arch of the foot and the ankle and permit even distribution of body weight.
- They should also protect the feet from being bruised against furniture, or by falling trays or utensils.
- Low- or flat-heeled shoes are usually the most suitable. They should be kept clean and used only at work to avoid germs from outside the premises being brought in.

## First-aid equipment

There should be an adequate and easily accessible supply of bandages, dressings and antiseptics on the premises for the first-aid treatment of workers who have accidents. Cuts or scalds should be covered with protective waterproof dressings so that any infection does not spread. Blue waterproof dressings are preferable, as they can be easily seen if they get into food.

At least one member of staff, trained in first aid, should be available whenever employees are on duty in the kitchen. The most convenient place for the first-aid kit is usually in the staff rest-room or cloakroom. Names of the staff trained in first aid and information on how to reach them quickly should be clearly displayed nearby.

### Personnel
### Important training points

- Remember, there is equal likelihood that food handlers will acquire bacterial contamination from the food they handle, as that they, as carriers of organisms, will contaminate food.
- Hand-washing, to remove contamination, should be repeatedly emphasized.
- As a matter of routine, management should look for obvious symptoms of any infection among staff that might be transmitted via food.
- Domestic pets should not be allowed in any food preparation or food storage area.

# Bibliography


DEPARTMENT OF HEALTH AND SOCIAL SECURITY. *Clean catering*, 4th ed. London, Her Majesty's Stationery Office, 1972.

LONGREE, K. & BLAKER, G. G. *Sanitary techniques in food service*, 2nd ed. New York, John Wiley & Sons, 1982.

NATIONAL INSTUTUTE FOR THE FOOD SERVICE INDUSTRY, *Applied foodservice sanitation*, 3rd ed., Boston, Little, Brown and Co., 1985.

RAJAGOPALAN, S. & SHIFFMAN, M. A. *Guide to simple sanitary measures for the control of enteric diseases*. Geneva, World Health Organization, 1974.

SALVATO, J. A. *Guide to sanitation in tourist establishments*. Geneva, World Health Organization, 1976.

UNITED STATES FOOD AND DRUG ADMINISTRATION *Food service sanitation manual*. Washington, DC, United States Food and Drug Administration, 1976.

WHO Technical Report Series, No. 785, 1989 (*Health surveillance and management procedures for food-handling personnel*: report of a WHO Consultation).

PART III

# SAFE FOOD HANDLING

Chapter 10
# Refrigeration

## Supervision of refrigeration

To ensure that refrigerators, cold rooms, and deep freezes are always in working order, kept clean and used efficiently and economically, a member of staff should be put in charge of the equipment. This staff member should be responsible for cleaning the equipment, and for deciding which food should be refrigerated, and how and where it should be stored. The refrigerator supervisor need not necessarily handle the food, but the number of people with access to refrigeration equipment should be kept to a minimum. The more people who have access to the equipment, the more frequently doors and lids will be opened and left open. Also if staff get into the habit of refrigerating food unnecessarily, the equipment will become too full, and less efficient.

### Refrigerators

A refrigerator stops bacteria from multiplying on, or in, food. The bacteria are not killed. Their growth is only arrested, and they will start to multiply again when the food is taken out of the refrigerator into a warmer environment. Refrigeration does not change the nature of the food itself. Food kept in a refrigerator will only remain in good condition for a limited period of time.

The temperature below which most foodborne pathogens stop multiplying is 10 °C.[1] Therefore, for the normal, short-term storage of perishable food, a temperature of 5–10 °C is recommended.

#### Packing a refrigerator

Priority for space in the refrigerator should be given according to Table 3.

To allocate each kind of food to the most appropriate position in the refrigerator, the supervisor needs to know where it is coldest and where the air circulates most effectively. It is advisable to consult the manufacturer for this information. Food should be placed in the refrigerator so that air can circulate freely around it. Poor ventilation and pockets of warm air can encourage the growth of mould and yeasts.

Generally liquids, for example gravies, should be placed close to the freezing coils. They should be stored in shallow containers, of not more than 5–8 cm in depth. Shallow containers give a larger cooling area for the liquid than deeper pans.

[1] See footnote, page 64.

Table 3. Allocation of space in a refrigerator.

| Store out of the refrigerator | Store in the refrigerator | Give priority for space in the refrigerator |
|---|---|---|
| Bottled foods | Butter | Cream |
| Bread | Cooked food not needed for immediate consumption | Custards |
| Cake | Cream | Gravy |
| Canned foods | Custard | Meats |
| Cereals | Eggs and egg dishes | Milk |
| Flour | Fish | Soup |
| Pasta | Game | |
| Pastry | Gravy | |
| Rice | Margarine | |
| Salt | Meat | |
| Sugar | Milk | |
| | Pastry containing cream | |
| | Poultry | |
| | Prepared meat dishes | |
| | Salads | |
| | Shellfish | |
| | Soup | |
| | Synthetic fillings | |

Domestic-type refrigerators can be adapted to provide space for shallow trays of liquid to be stacked one above the other. Meat products should be placed further away from the freezing coils, and eggs and other foods furthest away. Food that is to be stored for a long time needs to be in the coldest part of the refrigerator. Meat and fish that will shortly be needed for cooking should be kept in the less cold part, and food with a strong odour, for example fresh fish, should be kept as far away as possible from foods, such as butter, that absorb odours (taint) readily. Food in glass or metal containers should be kept in the lower part of the refrigerator, so that drips of condensation from the cold containers will not fall on to other food.

If possible raw food and cooked food should be stored separately to avoid cross-contamination risks. However, if this is not possible, remember always to store raw food, particularly meats, below cooked food (see Fig. 21). This reduces the cross-contamination risk of tissue fluid from the raw food dripping on to the cooked food.

Fig. 21. Cooked meats should be stored above raw meats in the cold room.

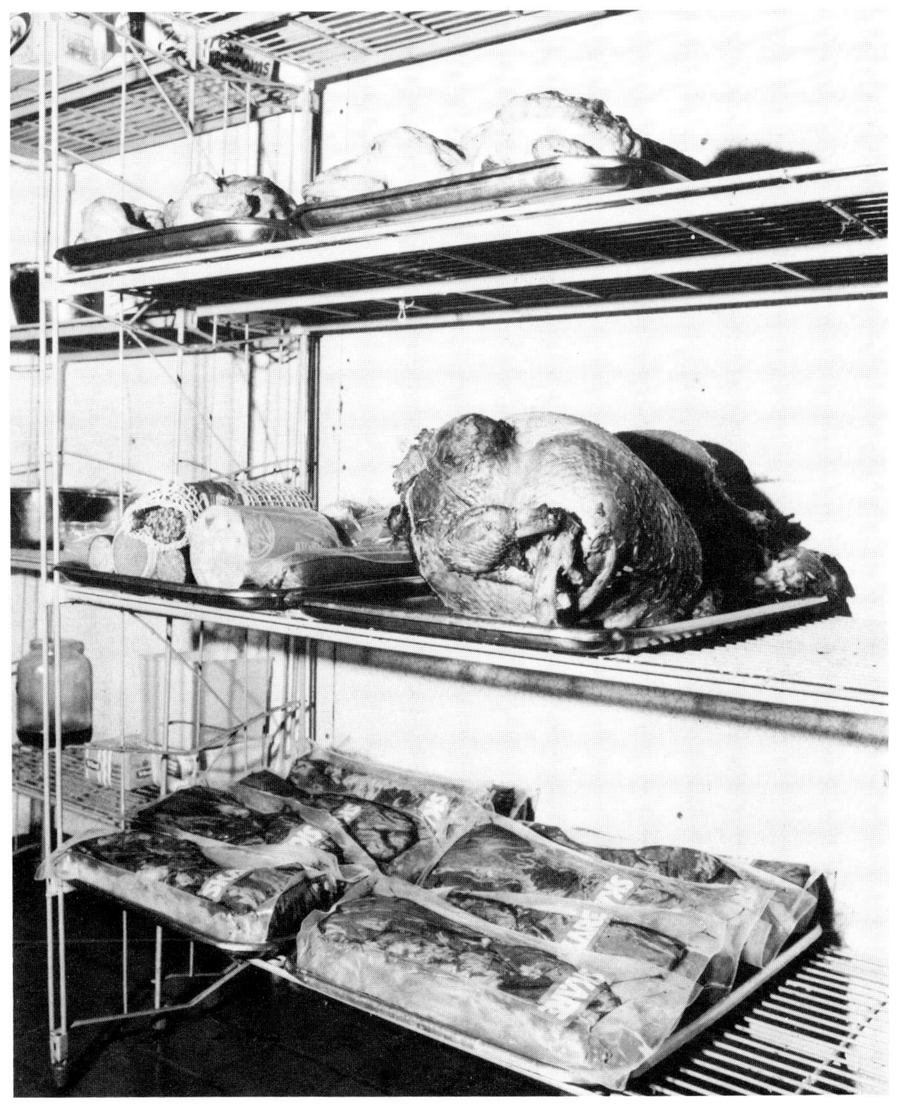

Photocentre, Eastbourne

## Refrigerator temperature checks

The temperature of refrigerators should be checked frequently. Supervisors should be equipped with probe or pocket, pencil-type thermometers to check the temperature of food in refrigerators, and to check the air temperature in cold rooms.

### Defrosting refrigerators

Defrosting must be carried out regularly as the cooling mechanism of a refrigerator cannot work properly, and may be damaged, if its coils become covered with a thick layer of ice. Exactly how often a refrigerator needs defrosting depends on how much it is used, the type and quantity of food it contains, and the heat and humidity of the food premises. However, once a week is usually often enough. Defrosting should be carried out when refrigerators are expected to be fairly empty. Careful planning may be necessary to achieve this.

Refrigerators must be kept clean, and the walls and trays should be washed during defrosting. The outside surfaces of refrigerators or refrigerated cabinets also need to be washed.

## Cold rooms

As much chill storage space as possible should be provided, especially in large establishments. A cold room, maintained within the temperature range of 3–4 °C, is advisable for cooling meat joints and poultry after cooking. As with refrigerators, the manufacturer should be consulted about the relative coldness of different parts of the room, and about air circulation, especially when deciding where to locate hooks and shelves.

Cold-room floors should be regularly scrubbed and mopped with a mild solution of disinfectant. Food should never be stored on the cold-room floor or below other food that could spill and cause contamination.

## Deep freezes

Few microorganisms are killed by freezing, so that storage of food at −18 °C cannot be relied upon as a means of destroying pathogens. For example, salmonellae have been isolated after years of frozen storage in whole egg products and meat. Spores of *Clostridium perfringens* and *Clostridium botulinum* have considerable resistance to alternate freezing and thawing at a temperature as low as −50 °C. Staphylococcal toxin has been shown to withstand a temperature of −18 °C for several months.

Freezing will not restore the freshness of food already contaminated or spoiled by bacterial action. When frozen food is thawed, the bacteria that have survived will start growing again and more spoilage will occur. The length of time that thawed food can be kept is limited, and it must not be left at room temperature for too long before being eaten.

A temporary period of thawing due to a cut in the electricity supply or a failure of the refrigeration mechanism does not necessarily mean that partially thawed food must be discarded. Provided the temperature of the food has not risen above 5 °C and the food is eaten without delay, it should cause no problems. However, frozen food that has been thawed should not be refrozen. This leads to deterioration in quality and to microbiological hazard. Discretion is needed when considering the use of partially thawed food. The decision will depend on the length of time the food has been thawing, the rise in temperature, and the general condition of the food. If the central core of the food remains frozen the outside will still be cold enough to stop most bacteria growing.

During food manufacture, and in large institutions, food is often frozen before being put into a deep freeze. This is most effectively achieved by blast freezing. The food is placed in a special chamber (the blast freezer) and subjected to a continuous blast of air at less than −20 °C. The length of time and the temperature needed vary according to the type of food. After freezing, food is packed into deep freezes and kept at −18 °C or below.

The shelf-life of frozen food varies according to the type of food, but in general food may be stored frozen for about 8 weeks, without any significant loss of nutrients or palatability. After that time foods with a high fat content may become rancid. Other foods can be satisfactorily stored for longer periods, and some almost indefinitely. A clear system of marking frozen food containers with a batch number and date should be operated so that stock can be rotated on the 'first in, first out' principle.

## Thawing

As a rule, food taken out of the deep freeze must first be thawed thoroughly, and then cooked immediately. Depending on its size, thawing of a turkey at room temperature takes between 8 and 12 hours. For larger birds and joints of meat of more than 10 kg in weight, a longer period may be necessary. Turkeys weighing 9 kg have been found to take more than 40 hours to thaw in a refrigerator, at less than 5 °C, and 9 hours at an ambient kitchen temperature of 25 °C. It is recommended that joints of meat be limited in size to 3 kg unless special care is taken to ensure thorough thawing.

If proper thawing is not achieved before cooking, the heat applied during cooking may not penetrate the whole carcass or joint, and bacteria may still be alive in the centre at the end of cooking. However, because tissue drip from thawing poultry and meat is a cross-contamination hazard, a different method can be employed. Frozen poultry and meat joints can be transferred straight from the deep freeze to the oven for cooking, without intermediate thawing at room temperature. Even though this differs from the general rule, it can be a safe practice provided that the cooking time is extended, or the temperature raised, to allow for initial thawing during the cooking process. Extreme care must always be taken when using this method.

## Cooling

Experiments to calculate the cooling rates of large joints of meat have shown that immediate storage in a well ventilated cooling room after cooking is the most effective cooling method.

In large extablishments it is advisable to have special cooling rooms where hot food can be exposed to a continuous current of cold air to cool it rapidly, before it is placed in a cold room or refrigerator. In smaller establishments a circulating fan installed in a well ventilated room will provide a satisfactory environment for cooling meats and poultry. In establishments where there is not enough space to have a separate room for cooling, hot meat and poultry should be left in a cool place with some air movement for up to $1\frac{1}{2}$ hours (no longer) before refrigeration.

## Refrigeration
## Important training points

- Ensure that there is an accurate thermometer in your refrigerator. Check the temperature of food stored in the refrigerator frequently. If refrigerators operate at a temperature higher than 10 °C, they may need defrosting or servicing.
- Poultry meat tends to drip a lot during thawing and afterwards, so it may be highly dangerous in terms of cross-contamination risks.
- Keep the inside of refrigerators clean, and wash the surfaces at frequent intervals.
- Cooked food must be stored separately from raw food. The most effective way to do this is in separate refrigerators.

Chapter 11

# Cooking

Fresh food, cooked and eaten while still hot, should never be the cause of foodborne illness. Even though many raw foods are contaminated with pathogenic bacteria when they arrive at a food service establishment, thorough cooking should kill the bacteria. However, if cooking is not thorough enough, bacteria will incubate within the food, and lead to foodborne illness. Some bacteria give rise to spores that can survive cooking. These spores will give rise to bacterial growth if cooked food is cooled too slowly, or stored at kitchen temperature for too long. Some foods need more careful cooking and storage than others.

Meat and poultry need the most care, especially if joints of meat or birds are large. Meat and poultry should be cooked so that an internal temperature of 70 °C is reached in the deepest part of the bird or joint. Chefs should be encouraged to carry out regular internal temperature checks of meat and poultry, using probe thermometers. Poultry meat has a high risk of being contaminated because bacteria can be present on the surface or in the abdominal cavity. Rolled meat joints need just as much care in cooking as solid joints. They carry the additional risk of the outer, and probably most contaminated, surface being rolled into the middle of the joint, where bacteria are most likely to incubate.

Shellfish are another high-risk food. They should always be cooked to a temperature of 70 °C. The shells may be heavily contaminated before cooking. If shellfish, in their shells, are to be added to a dish the shells must be thoroughly scrubbed beforehand. Cooked shellfish should never be cut up where raw fish is handled. If uncooked frozen shellfish are to be used, they should be boiled from frozen, and stored at no higher than 4 °C until needed. Any surplus after a meal should be discarded.

## Methods of cooking

Quick, high-temperature cooking is the best for food safety. These conditions are best achieved by methods such as steam under pressure, thorough roasting of small quantities of meat, grilling, and frying. Infrared rays and high frequency waves have good penetration characteristics, but may not distribute heat evenly through food.

### Convection ovens

In convection ovens, air is circulated by a fan to improve heat transfer, and to make cooking faster, more even, and more effective than in conventional ovens. As with conventional ovens, vegetative bacterial cells are killed, but not all spores.

### Microwave ovens

Microwave ovens reheat cooked foods very quickly. They are often used to reheat deep-frozen, cooked food for fast food service. Microwaves heat food by agitating the molecules, especially water molecules. Microwave cooking can have some disadvantages, for example, poor heat distribution, and lack of browning of meat joints. To overcome these problems, some microwave ovens have forced-air convection currents, and infrared heat sources so that they can roast, bake, fry and grill in the same way as conventional ovens. "Standing times" (i.e., the time that the food should be allowed to stand after being removed from the microwave oven) must be observed.

### Pressure steamers

Food can be cooked to order within minutes using pressure steamers. They create good conditions for the destruction of bacteria and spores by the combination of pressure and heat. In hard water areas it is usually necessary to install a water softening plant to treat the water to be used for steam generation.

### Slow cookers

Electrical slow cookers consist of glazed earthenware casseroles and lids, set into outer aluminium casings. The heat for cooking comes from an element wound between the earthenware casserole and the aluminium casing. It is essential that the manufacturers instructions for the use of slow cookers are followed rigidly, as cooking food at a low temperature for a long period of time is often hazardous. Food should be eaten hot, immediately after removal from the casserole.

## Hazardous techniques

### Undercooking

The degree to which meat is cooked is a matter of choice, but if it is not to be cooked thoroughly, right through, care must be taken to ensure that it is free from initial contamination. Heat penetrates meat slowly, so any meat that remains red inside has not reached the internal temperature of 70 °C needed to kill bacteria. Slowly cooked, rare beef has been the cause of many outbreaks of salmonellosis and *Clostridium perfringens* foodborne disease.

Other foods that may cause problems if inadequately cooked are unpasteurized milk, eggs and egg products, and gelatin. The recently highlighted increases in reported cases of *Salmonella enteritidis* infection in Europe serve to emphasize the hazards associated with inadequately cooked eggs and egg products.

## Reheating

In many food service establishments, large joints of meat or poultry are cooked, and then sliced ready to be reheated after a period of refrigeration or storage at ambient (room) temperature. This practice, although common, should be discouraged as it prolongs the time the meat is kept within a temperature range suitable for the multiplication of bacteria, particularly salmonellae and *Clostridium perfringens*. Fig. 22 gives some examples of how long it takes bacteria to multiply at different temperatures. To minimize the risk of foodborne illness, meat dishes should always be kept at over 60 °C or below 10 °C (i.e., either too hot or too cold for bacterial multiplication).

As a general rule it is safer to keep meat raw, and preferably cold, overnight and to cook it thoroughly on the day it is required. If reheating before serving is unavoidable, an internal temperature of 70 °C must be achieved. Temperature checks with probe thermometers should be routine.

Stews, soups, curries, minced meat dishes, gravies and sauces are common causes of foodborne illness. On reheating, the middle of the mass of food may not reach a high enough temperature to kill bacteria. The rule that such dishes should be cooked and eaten on the same day should be adhered to as strictly as possible, but if reheating is unavoidable they should be heated to 70 °C throughout, and maintained at that temperature for at least 2 minutes before being eaten.

Custard, and dishes containing custard, should be prepared and eaten on the same day if possible. If not, a reheating temperature of 70 °C must be achieved. Liquid or solid prepared dishes should never be reheated more than once, that is they should never be heated more than twice in all.

Fig.22. Rates of bacterial growth at different temperatures.

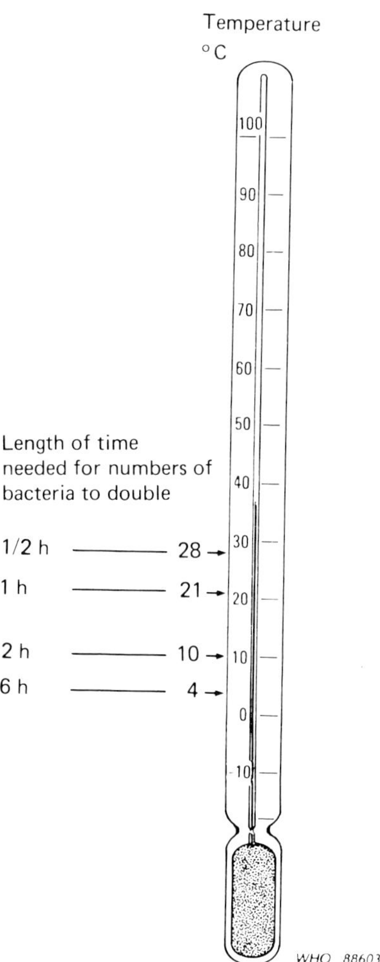

## Cooking
### Important training points

- Adequate cooking in the final stages of food preparation can eliminate earlier contamination.
- There are many routes by which food may acquire salmonellae. Thorough cooking at 70 °C, however, stops foodborne illness occurring.
- If you must reheat meat or meat dishes, the food should be brought to a temperature of 70 °C for at least 2 minutes.
- When meats, or dishes containing meat, are cooked and then stored for later consumption, the risk of foodborne illness immediately increases.
- Proper storage of food after cooking protects it from recontamination, and cold storage reduces bacterial growth.

Chapter 12

# Food preservation

Raw food is not sterile. All natural products contain, or carry on their surfaces, microorganisms that may cause either illness or deterioration and breakdown (spoilage) of the food. The bacteria that cause spoilage are quite distinct from those that cause foodborne illness. Spoilage bacteria are generally more resilient. They can sometimes multiply at lower temperatures than other bacteria, and can often withstand higher temperatures. For example, the combination of temperature and time used in the pasteurization of milk kills all pathogenic bacteria likely to be present, but allows spoilage bacteria to survive. If pasteurized milk is allowed to cool slowly, these bacteria will multiply rapidly as the milk passes through their preferred temperature range, and spoilage will occur. Ultra-heat-treated (UHT) milk, however, is treated at a sterilization temperature of 132–140 °C for 1–2 seconds. This temperature will kill all organisms present. UHT milk, in a sealed container, will keep unspoiled for months at room temperature.

In food processing and preservation, the aim is to eradicate both pathogenic and spoilage bacteria during manufacturing, so that food will remain safe to eat for a considerable length of time.

## Brining, curing and salting

The use of sodium chloride (salt) to preserve food has been common for centuries. Salt is used in brines and curing solutions, or is applied directly to food.

Salt produces an unsuitable environment for the growth of most bacteria. It dehydrates food by drawing out, and combining with, water molecules. This process produces chlorine ions that are harmful to bacteria. Salt also reduces the amount of oxygen that can dissolve in moisture in the food, stopping the reproduction of microorganisms that need oxygen. Salt also makes cells more sensitive to the harmful effects of carbon dioxide, and interferes with the action of enzymes.

## Canning

Some canned foods will be in storage and distribution for several years before they are eaten. Their condition should remain safe and acceptable if they are processed correctly at the time of canning, and not contaminated later. Normally cans are made of strong metal and can withstand wear and tear. The safety of canning depends on strict attention to detail. The raw food must be heated sufficiently to kill all pathogenic and spoilage bacteria, the cooling water must not be contaminated, and the seams of the can must form an adequate seal to prevent water and organisms from the outside getting into the can.

Never use food from rusty, dented, 'blown' or damaged cans, or canned food that smells or looks 'off' in any way. Food service establishments should not bottle or can their own food.

## Cook–freeze and cook–chill systems[1]

Cook–freeze and cook–chill systems are used widely in food service establishments, particularly in institutional catering. Cook–freeze systems are based on cooking, followed by fast freezing by plate or blast freezer, then storage at a controlled temperature of −18 °C, or below. The food should be reheated immediately before being eaten.

Cook–chill systems are based on the cooking of food, followed by fast chilling to, for example, a temperature of between 0 °C and +3 °C (i.e., above freezing point). Again, food should be reheated immediately before consumption. In cook–chill systems chilling should be started within 30 minutes of the food leaving the cooker, with a temperature of +3 °C being achieved within the next $1\frac{1}{2}$ hours. Shelf-life in the chiller is limited to 5 days, and reheating of food should raise the temperature to at least 74 °C. Temperature control of chillers is critical, and automatic temperature controls should always be used. A chiller used for a cook–chill food system should not be used for any other form of food storage.

Both cook–freeze and cook–chill systems have advantages for catering. Stores can be centralized, staff time saved, and portions controlled. However, in both systems, a mistake in preparation of batches of food to be stored could lead to a large-scale outbreak of foodborne disease. High quality foods must, therefore, be used and they must be cooked and reheated adequately.

## Chemical preservatives

Chemical additives can be used to preserve foods and help them retain their quality for long periods. Their constitution and

[1] Department of Health and Social Security *Pre-cooked frozen foods. Guidance on nutritional and hygienic implications produced by the Panel on Pre-cooked Frozen Foods, under the auspices of the Committee on Medical Aspects of Food Policy.* London, Department of Health and Social Security, 1970–1982.

quality must conform to the legal requirements of the country in which the food will be sold.

Commonly used preservatives include: sulfur dioxide in some meat products, fruit juices, coconut, and fruit; propionic acid in bread; ascorbic acid against mould growth in confectionery items, wines, cheeses and pickles; and sodium nitrate in cured meat and some cheeses.

## Drying and dehydration

Drying may be carried out in a number of ways. Sun-drying for fruits such as raisins, prunes, and figs is carried out in hot, sunny climates. Any method, for example salting, that reduces the amount of moisture available in food is a form of drying.

Mechanical dryers may be used for liquid foods such as milk, juices and soups. Evaporation of water is achieved at fairly low temperatures in a partial vacuum.

Other foods, for example meat, fish, coffee, tea, and eggs, can be dehydrated using heat and controlling conditions of relative humidity and airflow. Dehydration does not kill all micro-organisms, and the quality of the food will depend on the level of microbiological contamination before dehydration. Products should, therefore, be pasteurized before dehydration and filtered air should be used for the drying process. Packaging of dehydrated food to ensure low moisture content during storage will increase food safety.

## Freeze-drying

Freeze-drying is used for a number of foods including meats, poultry, seafood, fruits and vegetables. Drying is achieved by exposing quick-frozen foods to moderate temperatures in a vacuum.

## Irradiation

Irradiation is a physical method of food processing, comparable to methods such as heat treatment or freezing. It consists of exposing foods to gamma rays, X-rays or electrons over a limited period of time. The gamma rays used in food irradiation are mainly derived from cobalt-60. This is not a waste product of the nuclear industry, but is manufactured specifically for use in radiotherapy, the sterilization of medical products, and the irradiation of food.

The advantages of radiation over conventional food processing methods are that: foods can be treated after packaging; foods, for example meats, fish, and fruits and vegetables, can be kept in the fresh state; perishable foods can be kept for longer without loss of quality; and the cost and energy requirements of the process are lower than those of many conventional methods. Changes in the nutritional value of foods after irradiation are comparable with those produced by other methods of food preservation.

## Smoking

Smoking helps to preserve food and adds flavour. Chemical preservatives from the smoke impregnate the food, especially near the surface. The combined effect of these chemicals and the heating and drying effects of smoking preserve the food. The temperature, humidity, and duration of smoking needed depend on the type of food. Smoking temperatures for meats vary from 43 to 71 °C, and the smoking period from a few hours to several days.

Wood smoke is usually used, preferably from hard woods such as oak, maple, beech, birch, walnut and mahogany. Wood smoke is more effective against vegetative bacterial cells than against spores, and the rate of germicidal action depends on the concentration and temperature of smoke. Smoked foods are better protected against bacteria than against moulds.

## Sous vide

This system of food preparation involves the packing of prepared raw or partially cooked food into specially manufactured plastic bags, which are then evacuated and sealed. These are placed in a moist-heat oven to pasteurize the food. The bags are then cooled rapidly in a blast chiller, cryogenic chiller, or iced water. The food can then be stored at a temperature below 3 °C for up to 21 days, ready for re-heating and serving. Special care needs to be exercised to ensure that correct preparation and subsequent heating procedures are observed.

## Vacuum packing

Vacuum packing, or packing in the absence of oxygen, is a method of food preservation that is used especially for large quantities of cured meats—whole or sliced, cooked or uncooked—and for uncured cooked meats. The process preserves colour in both uncooked and cooked cured meats; prevents products from drying out; reduces spoilage by bacteria that need oxygen (aerobic bacteria) and prevents rancidity. At temperatures of 1–2 °C, cleanly butchered beef, for example, can be held in vacuum packs for up to 2 months.

The cut of meat to be vacuum packed is placed in a bag made of a substance with low permeability, air in the bag is drawn out, and the neck of the bag is sealed by heat or by being clipped. In most systems, heat-shrinkable films are used. After clipping or heat-sealing, the pack is either dipped briefly in hot water or exposed to hot air or steam in a shrink tunnel, which shrinks the bag over the meat, leaving minimal free space. Any residual oxygen is rapidly absorbed by the meat, and carbon dioxide is produced within the pack, inhibiting bacterial growth, and extending storage life at low temperatures. Vacuum-packed products should be kept refrigerated, coded with the date of manufacture and carefully rotated (i.e., older packs should always be used before newer ones).

Chapter 13

# Self-regulation and quality control

Self-regulation and quality control in food production are usually associated with manufacturers where the food produced is distributed nationally and internationally. The same principle should be applied to catering in hotels, restaurants and other food service establishments. Some form of self-inspection and control is needed for a variety of reasons.

## Quality control

It is essential that all food supplied to food service establishments comes from sound and reliable sources. Quality, in terms of wholesomeness and absence of contamination, should be ensured at the time of purchase. Food should also be checked for quality and absence of contamination when it arrives at the food premises. Depending on the size of the establishment, such checks may involve the occasional microbiological and chemical examination of samples of new products, or of new product lines.

Managers should be sufficiently trained and experienced to judge the quality and freshness of food by its look, feel, smell or taste, and to select samples for more detailed examination if fraud, adulteration, or contamination is suspected.

## Self-regulation

A customer's initial impression on entering a food service establishment often determines whether he or she will return. It is therefore very important that a customer should get an impression of a bright, clean and attractive eating place. The appearance and the personal hygiene practices of the employees are also very important, as is the way food is handled and the way equipment and utensils are cleaned and disinfected.

The customer's health is best protected if management is required to carry out continuous inspections of both their facilities and their practices. Checks should be made department by department, and function by function, for unsafe procedures and unhygienic conditions, that could lead to contamination of food and growth of harmful bacteria.

Self-inspections should cover the following:

- Personal cleanliness of food handlers.
- Food handling practices.

- Receiving points for raw food and food ingredients.
- Food storage facilities, including all refrigeration facilities.
- Food preparation areas.
- Food-holding equipment, ovens, warming cabinets, reheating cabinets.
- Dining-rooms and serving areas.
- Washing-up areas.
- Rest rooms, customers' cloakrooms, public facilities.
- Employees' facilities – toilets, locker rooms, lunch rooms.
- Storage facilities for supplies and equipment.
- Refuse collection and disposal areas.
- Boiler rooms, compressor installations, and other utilities.
- Entries, exits, and outside surroundings, including car parks, drive-in service areas, incinerators, etc.
- Vehicles used for transporting food.
- Any area that causes customer concern.

The local public health authority responsible for inspection will have laid down legislation for food safety for the areas mentioned above. Managers need to be familiar with this legislation for self-inspection programmes to be successful.

There may also be regulations that demand that illness among food handlers is reported to the local public health authority. Self-inspection should include arrangements for this notification.

The extent of self-inspection will depend on the scale of the catering operation. Large premises will need more detailed inspection than small snack bars or kiosks. However, the frequency of inspection should be based on the extent of hazards to health. Critical points where contamination risks may occur should be identified in methods and procedures, and in the use of equipment. These critical points can be identified using the Hazard Analysis Critical Control Point System (see page 110). The priority and frequency of inspection at these critical points will depend on the degree of contamination risk involved. Even the smallest establishment may be carrying out a catering process that incorporates a particular hazard to health, and requires specific self-inspection, for example the reheating of turkey meat, or the preparation of dishes containing cooked shellfish.

Self-inspection should primarily provide a continual check to ensure that supervision is adequate and complete, and to ensure the safety of all areas and activities of the food service establishment. If continual checks are made, they should help prepare for any inspection by the local public health authority. Depending on the size of the establishment, self-inspection reports could be compiled and maintained to allow for periodic review by management.

Inspections are needed of raw materials, finished products and all hygiene aspects of the establishment. Checks on facilities and equipment, daily supervision of cleaning and disinfection of equipment, the establishment of a personal hygiene programme for all workers, and the education of food handlers in basic food safety, are also required. Checks on water supplies, the efficiency of drainage systems, and the hygiene of vehicles used to supply food to the establishment, should also be part of self-inspection.

If a member of management is given responsibility for supervising the instruction of employees in food safety techniques, it is important that he or she has the technical knowledge required and is able to communicate successfully with employees. He or she may require special training in the teaching techniques required. Part IV of these guidelines gives details of training methods and outlines a sample training programme.

If an outbreak of foodborne illness occurs in the food service establishment, either among employees or customers, a management nominee should have specific responsibility for contacting the local public health authority. He or she should ensure that every necessary facility is provided for the officers who come to investigate the outbreak.

In larger food service establishments, using sophisticated catering systems, for example microwave ovens and vending machines, or cook–freeze and cook–chill systems, it may be helpful to have access to a laboratory where food can be tested for safety and quality. Any food that indicates a safety risk, poor quality, or is otherwise unlikely to meet consumer expectations, should be rejected.

## The Hazard Analysis Critical Control Point (HACCP) system[1]

This system is emerging as the primary means by which the food industry can ensure the safety and quality of its products.

[1] ICMSF. *Application of the hazard analysis critical control point (HACCP) system to ensure microbiological safety and quality.* Oxford, Blackwell Scientific Publications, 1988 (Microorganisms in food, No. 4).

Hazard analysis assesses the risks associated with the various operations of food production, for example growing, harvesting, processing, manufacturing, distributing, marketing, and preparing, to identify those in which control is essential to ensure food safety and quality.

The HACCP system is equally applicable in food service establishments in developing and developed countries, and consists of the following elements.

1. Hazard analysis. The identification of **hazards**, followed by assessment of the **severity** of these hazards, and the **risks** that they pose.

A **hazard** is the unacceptable contamination, growth, or survival in food of bacteria that may affect food safety or quality and/or the unacceptable production or persistence in foods or products of substances such as toxins, enzymes, or products of microbial metabolism.
**Severity** is the magnitude of the hazard or the seriousness of the consequences that can result when a hazard exists.
**Risk** is an estimate of the probability of a hazard occurring.

2. Determination of **critical control points** required to prevent or control the identified hazards.

A **critical control point** is an operation (practice, procedure or process), or a step in an operation, in which a preventive or control measure could be exercised to eliminate, prevent, or minimize a hazard.

3. Establishment of effective preventive or control measures and specification of **criteria** that indicate whether an operation is under control at a particular critical control point.

**Criteria** are specified limits or characteristics of a physical (for example, time or temperature), chemical or biological nature, which ensure that a product is safe, and of acceptable quality.

4. **Monitoring** of each critical control point to evaluate whether it is under control.

**Monitoring** is checking that the processing or handling procedure at each critical control point meets the established criteria. It involves systematic observation, measurement, and/or recording of the significant factors for prevention or control of the hazard. The monitoring procedures chosen must enable action to be taken to rectify an out-of-control situation or to bring the product back into acceptable limits, either before or during an operation.

5. Implementation of appropriate and immediate corrective action whenever the monitoring results show that criteria established for safety and quality at a critical control point are not met, i.e., that the **operation** is out of control.

In this context an **operation** is regarded as having ended when the responsibility for a batch of materials or a food product, or the material or product itself, has changed hands, or when the food has been eaten.

To evaluate whether the HACCP system is functioning as planned, supplementary tests are used, or records reviewed. It is necessary to test whether appropriate critical control points have been identified, that they are being effectively and properly monitored, and that appropriate action is being taken whenever criteria do not fall within specified limits.

## Associations of hotels and restaurants

In large centres of population, holiday resorts or conference centres where there are several hotels, restaurants and other public eating places, it can be beneficial for the managers of the businesses to form an association to promote and maintain standards. The association can devise its own rules and standards to provide a way of applying quality control to the businesses belonging to the association. The association can also set up training courses or on-the-job training schemes and ensure that staff of businesses belonging to the association are properly trained.

In areas where the government, local government or a tourist agency promotes the hotel and restaurant trade, and publishes guides and publicity material, promotion could be restricted to premises that conform to the standards required for association membership. The government, local government or tourist agency could employ staff to inspect premises to ensure that these standards are being adhered to, before new publicity and promotional material is produced.

Self-regulation and quality control in food service establishments can, therefore, be incorporated into a public policy of setting a high standard for the industry, and can help to promote tourism, conference facilities and general trade in local areas.

## The foodborne disease hazard league table and the food safety hazard check-list

Investigations into the causes of 1000 outbreaks of foodborne illness have shown that certain identifiable factors occur in all outbreaks. Some outbreaks are caused by the combined effect of five or six of these factors, whereas others are caused by only one.

The factors (hazards) can be listed into a 'league table' (see page 114), according to the likelihood of their causing foodborne illness. The first 10 hazards from the league table are the most likely to cause foodborne illness. They can be arranged into a hazard check-list (see pages 115–117) that can be used by management carrying out self-regulation in food service establishments. For each hazard, or group of hazards, on the check-list a number of points that must be considered by management are listed.

Table 4. The foodborne disease hazard league table[1]

1. Food prepared too far in advance.
2. Food stored at room temperature.
3. Cooling food too slowly prior to refrigeration.
4. Not reheating food to a high enough temperature to destroy pathogenic bacteria.
5. The use of cooked food that was contaminated with pathogenic bacteria before it was cooked.
6. Undercooking meat and meat products.
7. Not thawing out frozen meat and poultry for long enough.
8. Cross-contamination from raw to cooked food.
9. Storing hot food below 63 °C.
10. Infected food handlers.

- - - - - - - - - - - - - - - - - - - - - - - - - - - - - - - - - - - - - -

11. The use of left-over food.
12. Eating raw food.
13. Preparing extra-large quantities of food.
14. Contaminated canned food.

[1] Adapted from: Roberts, D. Factors contributing to outbreaks of food poisoning in England and Wales 1970–79. *Journal of hygiene,* **89**: 491–498 (1982).

Fig.23. Food hygiene hazard check-list.

HAZARDS 1, 2 & 9

FOOD PREPARED TOO FAR IN ADVANCE
FOOD STORED AT ROOM TEMPERATURE
STORING HOT FOOD BELOW 63°C

A. Check for perishable cooked food on arrival in delivery containers. Check temperature on arrival, storage temperature—is it under 10°C? (Perishable food to note particularly—ham, tongue, cold meats, meat pieces, dairy products, seafood, egg products, mayonnaise, gelatin, glazes, powder mixes.)
B. Are there any meat or poultry dishes at ambient temperature in the kitchen that will not be consumed within 2 hours?
C. Is any food stored on heated trolleys or are there any hot storage cabinets or other hot storage facilities holding food under 63°C.
D. Is there any minced meat or sandwiches, prepared and waiting to be served?
E. Are any leftovers from meals awaiting a decision on further use?
F. Is there any refigerator breakdown?
G. Is there any means of measuring the temperature of food in the refrigerator?

HAZARD 3

COOLING FOOD TOO SLOWLY PRIOR TO REFRIGERATION

A. To what extent are joints cooked for subsequent slicing and reheating?
B. Is there a separate cooling room or other cooling equipment or an area of preparation space for cooling joints and prepared meat dishes?
C. Stews, sauces and stocks, minces, curries, gravies—what is the method of preparation, how often are they prepared and how long are they stored prior to consumption?
D. What refrigeration space is available? Is there enough capacity for foods requiring cold storage after ambient cooling?

HAZARD 4

NOT REHEATING FOOD TO HIGH ENOUGH TEMPERATURE TO DESTROY PATHOGENIC BACTERIA

A. Is there any means of measuring the temperature of meat and poultry during cooking?
B. How are foods reheated—steamed, boiled, microwave or normal oven?
C. Are cold meats reheated by addition of hot gravy/sauces?

HAZARD 6

UNDERCOOKING MEAT AND MEAT PRODUCTS

A. Is there any intentional undercooking of meat?
B. Are there any special slow-cooking appliances?

HAZARD 7

NOT THAWING OUT FROZEN MEAT AND POULTRY FOR LONG ENOUGH

A. Is the period allowed for thawing out related to the size of the joint?
B. Is there any cooking direct from the frozen state?
C. Are joints immersed in potable water to thaw out?
D. Is any maximum size of joint employed, any portioning?
E. How and where is thawing usually done?

HAZARDS 5 & 8

THE USE OF COOKED FOOD, CONTAMINATED WITH PATHOGENIC BACTERIA BEFORE IT WAS COOKED
CROSS-CONTAMINATION FROM RAW TO COOKED FOOD

A. Meat and poultry supplies—what are the sources? Is there any knowledge of previous salmonella contamination incidents?
B. Areas where meat and poultry joints are thawed and prepared—is there separation from other foods, particularly cooked foods?
C. Is there separate refrigeration storage for raw and cooked meat?
D. Are delivery containers reused for other foods. Is there adequate separation of raw and cooked meat containers?
E. To what extent do food handling personnel cross from raw preparation to cooked food areas?
F. Butchery area—are cutting surfaces in good condition? Are they regularly cleaned? What are the cleaning procedures?
G. Provision of hand wash basins, hot water and soap etc. in food preparation areas—are the facilities clean and obviously used for hand-washing?
H. Staff protective clothing—is it in all respects adequate and clean?
I. Are cleaning schedules adhered to?

HAZARD 10

INFECTED FOOD HANDLERS

A. What food hygiene training have supervisory/food handling staff received?
B. Is there any history of foodborne illness among food handlers?
C. What supervision of infection risks are carried out—hand inspection—checks on symptoms of foodborne disease—reports of absences or holidays abroad?
D. Are first aid facilities provided?
E. Is a copy of the food hygiene legislation provided?

Adapted from: Roberts, D. Factors contributing to outbreaks of food poisoning in England and Wales 1970–79. *Journal of hygiene,* **89,** 491–498 (1982).

Chapter 14

# Summary

## The essentials of food safety

- Food must be adequately treated to inhibit the growth of, or to destroy, any microorganisms present that may cause disease.
- Recontamination of food must be avoided.

These essentials are achieved by a combination of elements outlined in this guide.

### A clean environment

Chapter 8 outlined how a clean environment can be achieved and maintained for the storing, handling and serving of food.

### Conscientious handling

Chapter 9 outlined the importance of personal hygiene and careful treatment of food by food handlers.

### Refrigeration

Chapter 10 outlined the importance of refrigeration for maintaining a cool or cold environment to inhibit the growth of bacteria.

### Cooking

Chapter 11 spelt out the importance of adequate cooking to destroy any pathogenic organisms in food.

# Bibliography


ASTON, G. & TIFFNEY, J. *A guide to improving food hygiene*. London, Northwood Publications, 1981 (Available from Eaton Publications, Walton-on-Thames, England).

BRYAN, F. L. & MCKINLEY, T. W. Control of thawing, cooking, chilling and reheating turkeys in school lunch kitchens. *Journal of milk and food technology*, **37**: 420–429 (1974).

CHRISTIE, A. B. & CHRISTIE, M. C. *Food hygiene and food hazards*, 2nd ed. London, Faber and Faber, 1977.

DEPARTMENT OF HEALTH AND SOCIAL SECURITY. *Guidelines on pre-cooked chilled foods*. London, Her Majesty's Stationery Office, 1980.

FOOD AND AGRICULTURE ORGANIZATION OF THE UNITED NATIONS. *Manuals of food quality control. 5. Food inspection*. Rome, FAO/UNEP/WHO, 1984.

*Food irradiation: a technique for preserving and improving the safety of food*. Geneva, World Health Organization, 1988.

HOBBS, B. C. & GILBERT, R. J. *Food poisoning and food hygiene*, 4th ed. London, Edward Arnold, 1978.

NATIONAL RESTAURANT ASSOCIATION. *A self inspection programme for food service operators on sanitation and safe food handling*. Chicago, National Restaurant Association, 1973.

WILSON, N. R. P., ed. *Factors affecting quality control*. Barking, Applied Science Publishers, 1981.

PART IV

# TRAINING

Chapter 15

# Planning and implementing a training programme

Using the information in this guide, managers in food service establishments should arrange organized training sessions for their staff. The training should aim to achieve the following.

- It should encourage employees to be responsible and conscientious in providing safe, high quality meals that customers are satisfied with.
- It should provide employees with a knowledge of the hygiene practices they should adopt, and of any legal requirements they must conform to.

Communication with staff in the food industry often presents a problem, for two main reasons. The first is that many employees at the 'lower levels' of the food industry have little capacity or motivation to change their working habits. Many of them have jobs that are physically hard, and involve working long hours in uncomfortable conditions, for relatively small financial reward. Many workers are transitory. If pressed they will move elsewhere to find conditions where fewer demands are made. Many of the workers will have had little formal education, and there may also be a language problem. Hotel and restaurant kitchens often employ workers from different countries whose levels of literacy, even in their mother tongue, vary greatly. Communication between management and employees may therefore be limited to the spoken word, and this may not be fully understood.

Managers must be aware of these obstacles to educating food handlers. With appropriate knowledge, enthusiasm and persistence on the part of management, habits can be changed and standards improved, despite the difficulties. Often in the catering industry small groups of people work together, for example, in a hotel or restaurant kitchen. These workers are fairly isolated from other groups, and their working conditions are often specific to their group. Health educators therefore have to make a lot of isolated efforts, and have to be extremely adaptable to put across their health education messages. To achieve the best results it is suggested that four basic rules are followed.

## Teaching methods

'It is important that all teaching should be flexible and should have local relevance. The object of teaching is not only to tell people what to do, but to tell them why they need to do it, so that they themselves will be continually motivated to maintain the proper hygiene standards. A teaching programme should win the interest of the students; it should not challenge their understanding'.[1]

### 1. Keep it short

This is the essence of teaching. People with limited education may not be able to give their full attention for more than 10–15 minutes at any one time. Food hygiene education therefore should be given in frequent short bursts.

### 2. Keep it simple

Use simple language that is easily understood. Whenever possible avoid scientific or technical terms. For example, while it may be useful to mention the terms pathogen and bacteria so that people will be familiar with them, afterwards the term 'harmful germs' would do just as well and would be easier to understand. Educators must remember that their aim is to make themselves understood, rather than to display the extent of their own knowledge.

### 3. Keep groups small

Small groups of between one and four people are much easier to teach and influence than larger groups. In a small group, those listening are more likely to ask questions. It often requires a lot of courage for an employee to admit, in front of many colleagues, that he or she has not understood something.

### 4. Teach by example

If lack of hygiene and poor standards are accepted by the management, they will be adopted as the norm by employees. Teaching by example will demonstrate what is right, and will set the standards for employees to follow.

[1] CHARLES, R. H. G. *Mass catering*. Copenhagen, WHO Regional Office for Europe, 1983 (European Series No. 15).

## Planning training

If training is to be effective, it must be planned. Management should not wait for an outbreak of foodborne illness to occur in their premises before they decide to train their staff, nor should training be used to fill in time when the staff have nothing else to do. Adequate time and resources should be properly allocated for training. The time spent on training and the progress of each member of the staff should be recorded. In this way it will be possible to establish which of the employees require further training or retraining.

When planning a training session, it is useful to ask yourself the following questions.

### 1. About the group

- Who are the trainees?
- What is their age range?
- What is their experience in general, and in relation to the subject of the session?
- What are their interests?

### 2. About the objective of the session

- What is the objective of the session?
- What should the trainee know, or be able to do, at the end of the session?

### 3. About visual aids and other equipment

- What visual aids will be used during the session?
- What other equipment and materials will be needed, for example paper, pencils, etc?

Instant success should not be expected in teaching food hygiene. To alter people's attitudes is an uphill struggle at the best of times. Legislation and discipline can together bring about superficial compliance with hygiene standards, but the really dependable and worthwhile alteration in habits will come from a desire to apply hygienic methods. The only way food handlers can develop such a desire is through health education.

Communication of this education is helped by asking open-ended questions that require answers from the group. For example, questions beginning with 'how, why, what, when, who, where and which' usually stimulate a response. For each session the educator should plan some of these questions in advance.

## Teaching aids

There are many teaching aids and resources available to make the job of a health educator easier.[1] These include film strips, slide shows, videos, display boards, posters, games, and competitions. In talks, overhead projectors can be used with ready prepared transparencies to illustrate points. If the educator is skilled in the method, he or she can write directly on to the transparency during the session.

All of the above teaching aids are useful if used in an organized and logical manner. Managers responsible for teaching may not be experienced in the art of communication. It is therefore important to consider which teaching method is most suitable for the occasion. This may only be determined by trial and error, but in any attempt to get a message across, the illustration of a point on a slide or transparency may make it much clearer than would a verbal explanation alone.

Posters are very effective for passing on instant messages, both during education and afterwards, to remind food handlers of their legal and general obligations. It is beneficial if posters can be changed at intervals to vary the message, or to present the same message in a different way.

The main advantages and disadvantages of using visual aids (films, slides, videos, transparencies, and posters) may be summarized as follows.

### Advantages

- Visual aids break up presentations and provide an alternative means of learning.
- They can be used to show actual situations or examples.

### Disadvantages

- The information on the transparencies, etc. may become dated.

[1] BATES, D. *Food Safety. An international source list of audiovisual materials.* Prepared for the WHO Food Safety Programme by the British Life Assurance Trust for Health and Medical Education, London, 1987. Unpublished WHO document EHE/FOS/87.1.

- There is normally less chance for participation by the group during a film or slide show, than during a straightforward talk or group discussion.

### Booklets and leaflets

Leaflets or booklets can be handed out to workers, perhaps with their weekly or monthly pay slips, as part of a special food safety programme, or in an attempt to put over a routine message. The exact message will vary from country to country, and with the type of food service operation.

Carefully selecting the information to be put in a leaflet, and making a number of points gradually over a period of time, can be of great benefit in letting new employees know about the responsibilities they will be taking on in their jobs, and about the food safety habits they should develop.

## Talks

The information, figures, and 'Important Training Points' given in Parts I–III of this guide can be used by management as a reminder of the facts they should know, and as a basis for training food handlers.

It is important to provide the correct amount of training for food handlers, that is, an amount consistent with the average learning ability of the group.

A suggested training schedule is outlined below. It lasts for a total of 7 hours, and can be divided into 14 half-hour sessions, or 28 quarter-hour sessions, depending on the timespan of attention of the food handlers. The sessions could be held once or twice a week, or spread over a longer or shorter period. Updating or revision sessions can be held at a later date. It is important not to try to teach too much at one time.

## Sample training schedule

If half-hour sessions are chosen, 14 talks could be organized using the information given in this guide. During a talk it is important to encourage participation and feedback from members of the group.

### Session 1 — Introduction

Foodborne illness — outlining the problem. The effect of foodborne illness on digestion. Introduction to microbiology.

PART I CAUSES OF FOOD CONTAMINATION

**Session 2 — Bacteria**
Bacterial growth. Bacterial spores. Bacterial toxins. Sampling for the presence of bacteria. How bacteria cause foodborne illness. Bacterial foodborne diseases. Other communicable bacterial diseases that may be foodborne.

**Session 3 — Other food contaminants**
Viruses. Chemicals. Parasites. Natural food contaminants.

**Session 4 — Incidents of foodborne illness**
Case studies.

**Session 5 — Sources and transmission of food contaminants**
Meat. Seafood. Eggshells. Pets and other animals. Insects. The soil. The human body. Animal foodstuffs.

PART II PREVENTION OF FOOD CONTAMINATION

**Session 6 — Structure and layout of the food premises**
The kitchen. Working areas. Washing-up rooms. Crockery store. Toilets and cloakrooms. Dining-rooms.

**Session 7 — Equipment**
Crockery, cutlery, pots and pans. Surfaces. Sinks. Refrigerators. Vending machines. Microwave ovens.

**Session 8 — Maintaining a clean environment**
Detergents and disinfectants. Refuse disposal. Pest control.

**Session 9 — Personnel**
Health surveillance. Personal hygiene. Protective clothing.

PART III SAFE FOOD HANDLING

**Session 10 — Refrigeration**
Supervision of refrigeration. Refrigerators. Cold rooms. Deep freezers. Thawing. Cooling.

**Session 11 — Cooking**
Methods of cooking. Hazardous techniques.

**Session 12 — Food preservation**
Brining, curing and salting. Canning. Chemical preservatives. Cook–freeze and cook–chill systems. Drying and dehydration. Freeze-dying. Irradiation. Smoking. Vacuum packing.

### Session 13—Self-regulation and quality control

The Hazard Analysis Critical Control Point (HACCP) system. Associations of hotels and restaurants. The foodborne disease hazard league table and the food hygiene hazard check-list.

### Session 14—Summary

The essentials of food safety. Course assessment.

## Assessment and evaluation

Most of the last session can be used to evaluate the course, by assessing how much information the food handlers have absorbed. If any food handlers have been unable or unwilling to learn, the educator needs to think of ways to adapt the training to suit food handlers' needs better. For example, adding a few practical demonstrations may make points clearer, and the sessions more entertaining.

## Assessment tests

These can be either oral or written. If the facility exists, food handlers could sit a written examination organized by an independent health institute or agency. Food handlers may find it satisfying, and useful when seeking employment, to gain some sort of formal qualification.

## Oral questions

Examples of simple questions that could be asked orally are as follows:

1. What are pathogenic bacteria?
2. At what temperature do pathogenic bacteria generally multiply at their fastest rate?
3. What is pasteurization?
4. Why are refrigerators useful in the preservation of perishable foods?
5. How is food most frequently contaminated during its preparation?
6. Why should food handlers wear head coverings?
7. What types of food provide the best medium for bacterial growth?
8. Why should food handlers wash their hands before preparing food?

9. What dangers are associated with an infestation of rodents in food premises?
10. Which common insect is most likely to transmit pathogenic organisms to open food?
11. A raw chicken is placed on a tray on which cooked chickens are cooled. What is the danger?
12. How should products containing meat or cream be stored overnight?
13. Do pathogenic bacteria always produce signs of decay in food?
14. What is a healthy carrier?
15. Which illnesses, infections or conditions among food handlers should be reported?

## Multiple choice questions

Other questions, either as an alternative or an addition to the oral test questions above, may be put in the form of multiple choice. Food handlers should be instructed to circle A, B, C or D. Answers to the multiple choice questions are given on page 132.

1. Does refrigeration kill all dangerous bacteria that may be present in food?
   - A No, but it keeps foods cool so that bacteria do not multiply.
   - B It kills some bacteria that are susceptible to low temperatures.
   - C It kills all of them.
   - D On the contrary, it promotes growth.

2. Why do you think it is important to wash your hands before handling food?
   - A To make them look clean and attractive.
   - B Because the law requires it.
   - C To reduce the risks of germs on the hands contaminating the food.
   - D To reduce the risk of the food handler becoming contaminated from the food.

3. What is the safest temperature for storing food containing meat, eggs, or milk?
   A 5 °C
   B 10 °C
   C 15 °C
   D 38 °C

4. Which type of surface is the most hygienic for cutting up and preparing food?
   A Wood block.
   B Hardened rubber compounds.
   C Hardwood cutting boards.
   D Hard laminated plastics.

5. What is the best method of controlling flies?
   A Elimination of breeding places.
   B Destruction by spraying with insecticides.
   C Prevention of access to food.
   D Fly electrocutors.

6. A food handler with a cut finger poses a risk of causing foodborne illness, particularly from the organisms of:
   A Salmonella.
   B *Clostridium perfringens.*
   C *Bacillus cereus.*
   D *Staphylococcus aureus.*

7. In which temperature range is it generally unsafe to hold cooked meat?
   A −18–0 °C
   B 0–5 °C
   C 63–68 °C
   D 10–60 °C

## Essay questions

Employees with more advanced knowledge, or a higher standard of education, may find it more rewarding to answer essay questions. Some examples of questions that require essay-type answers are given below.

The educator needs to decide how long the employees should spend on each answer. For example they could be asked to answer three of the five questions in an hour, or they could be asked to answer all the questions, but not to write more than two short paragraphs on each.

1. Why is food safety education desirable for employees in food service establishments?
2. What are the dangers of reheating previously cooked meat?
3. What do you understand by the term cross-contamination?
4. Give examples of perishable foods and the ways in which they should be stored in food service establishments?
5. What do you understand by the term 'healthy carrier'?

### *Multiple choice answers*

1. A, 2. C, 3. A, 4. D, 5. D, 6. D, 7. D.

# Bibliography

BATES, D. *Food Safety. An international source list of audiovisual materials.* Prepared for the WHO Food Safety Programme by the British Life Assurance Trust for Health and Medical Education, London, 1987. Unpublished WHO document EHE/FOS/87.1.

SPRENGER, R. A. *Hygiene for management. A text for food hygiene courses.* Rotherham, Highfield Publications, 1986

# Index